trim
made simple

trim made simple

Book & DVD guide

Gary Katz

The Taunton Press

To the memory of Steve Phipps, friend and brother in the craft.
To Cynthia Cook, who taught me the importance of teaching.

Text © 2009 by Gary Katz
Photographs © 2009 by Gary Katz
Illustrations © 2009 by The Taunton Press, Inc.
All rights reserved.

The Taunton Press, Inc., 63 South Main Street, PO Box 5506, Newtown, CT 06470-5506
e-mail: tp@taunton.com

Editor: Steve Culpepper
Copy editor: Candace B. Levy
Indexer: Jay Kreider
Jacket/Cover design: Loop Design, Jason Carreiro
Interior design: Carol Petro, Chika Azuma
Layout: Carol Petro
Illustrator: Mario Ferro
Photographer: Gary Katz
DVD Production: Onsite Productions
DVD Editing: The Dream Tree

Library of Congress Cataloging-in-Publication Data
Katz, Gary, 1952-
 Trim made simple / Gary Katz.
    p. cm.
 ISBN 978-1-60085-054-7
 1. Trim carpentry--Handbooks, manuals, etc. I. Title.
 TH5695.K38 2009
 694'.6--dc22
                         2008022513

Printed in the United States of America
10 9 8 7 6 5 4 3 2 1

The following manufacturers/names appearing in *Trim Made Simple* are trademarks: DAP®, Multimaster®, Nail Gripper®, Phillips®, Sharpie®, Starrett®, Starrett® Prosite™, Super Glue®, Thumb Saver®, Trim Gauge®, True Angle®

Homebuilding is inherently dangerous. Using hand or power tools improperly or ignoring safety practices can lead to permanent injury or even death. Don't try to perform operations you learn about here (or elsewhere) unless you're certain they are safe for you. If something about an operation doesn't feel right, don't do it. Look for another way and keep safety foremost in your mind when you're working.

acknowledgments

This book wouldn't have been possible without help from countless carpenters and friends who have taught me techniques and added invaluably to the meager tool chest of what I've learned on my own, beginning with my brother—Larry Katz. I also want to thank Jed Dixon, Mike Sloggatt, Greg Burnet, Jim Chestnut, David Collins, Daniel Parish, and for all the carpenters I've sadly forgotten to mention, I blame my age.

My deep appreciation to Steve Culpepper, who proved himself a sensitive, understanding editor.

For readers who wish to pursue finish carpentry further, please visit my website: GaryMKatz.com

contents

Chapter 1: Make a Workstation	2	**Chapter 3: Windows**	34
Why You Need a Saw Stand	5	Cutting and Installing the Stool	35
Building the Workstation Step-by-Step	6	Cutting and Installing the Casing	38
		Casing a Window Step-by-Step	39
Chapter 2: Doors	16		
Cutting Casing	19	**Chapter 4: Baseboard**	52
Preparation	20	Planning Baseboard Joinery	52
Installation	21	Cutting Baseboard	55
Using Hand-Driven Nails	22	Installation	56
Trimming a Door Step-by-Step	24	Installing Baseboard Step-by-Step	58

 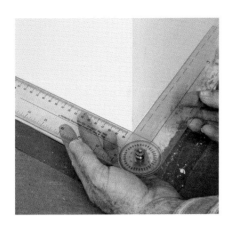  

Chapter 5: Chair Rail	70		**Resources**	103
Planning Chair Rail	71			
Cutting the Parts	73		**Index**	104
Installation	74			
Installing Chair Rail Step-by-Step	76			
Chapter 6: Crown Molding	84			
Cutting Crown Molding	85			
Measurement and Layout	87			
Cutting Accurately	88			
Installation	88			
Installing Crown Molding Step-by-Step	90			

CHAPTER 1

Make a workstation

Trim carpentry depends almost entirely on cutting clean, tight miters at precise angles using precise measurements. You can cut miters in most small moldings using a miter box and handsaw; but for large profiles, especially tall baseboard and crown molding, a power miter saw is the only way to go. Because power miter saws are now so affordable, anyone with an interest in carpentry should own one. If you're changing the moldings in your home, at the very least, consider renting one.

But there's no need to drain your savings account for the best saw. No matter how much or how little you invest in a miter saw, the quality and enjoyment of your work will depend as much on your saw stand as on the miter saw itself.

A miter saw workstation plan

For secure and safe cutting, ease of making repetitive cuts, and basic efficiency, a miter saw workstation goes a long way toward making your trim projects go more smoothly—without much extra cost.

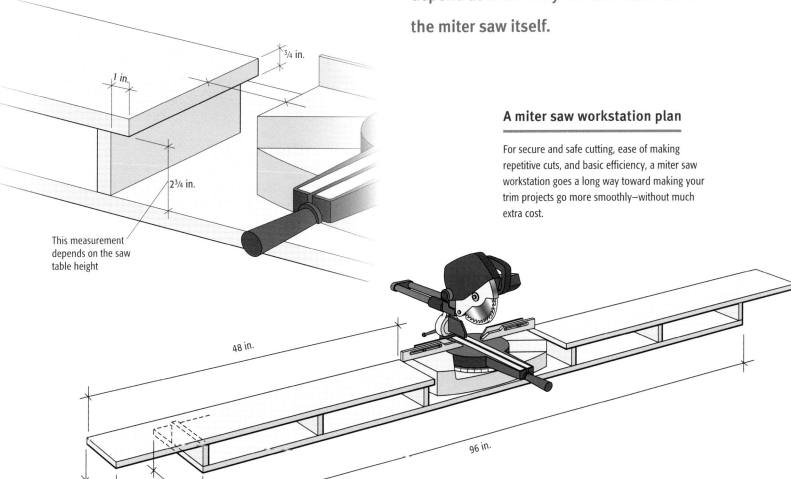

TOOL BOX

In this chapter, I begin to help you collect the tools you'll need for finish work. To guarantee success, be sure to have the right tools before starting the work shown in each chapter. By the time you work through the book, you'll have all the tools you'll need for almost any trim job.

Tape measure

Cordless drill

Countersink bit, Phillips® driver, and square driver

Clamps

Pocket hole jig

Miter saw

Carpenter's square

TOOLS AND MATERIALS

- Tape measure for measuring (a stiff 1-in. by 25-ft. tape is best for finish work)

- Cordless drill for drilling holes and driving screws

- Countersink bit, Phillips driver, and square driver for drilling countersunk holes and driving screws

- Clamps for securing material while working with tools

- Pocket hole jig for cutting pocket holes—the fastest method for precise joinery

- Miter saw for cutting moldings and millwork

- Carpenter's square for marking and measuring boards and trim

- 1-in. by 16-in. by 8-in. pine, fir, or plywood board, for the base of the miter saw stand

- 1-in. by 12-in. by 8-in. top pine, fir, or plywood for the top extension wings on the miter saw stand*

- 1-in. by 4-in. by 8-in pine or fir supports, ripped to the exact height of your miter saw (minus $3/4$ in.)

* 1 in. dimensional lumber is actually $3/4$ in. Plywood sold as $3/4$ in. is often thinner. Check the exact dimensions.

Why you need a saw stand

A miter saw stand is more than just a place to set your saw—it's a workstation. The stand must have continuous extension wings, so you can support different lengths of material. It must have a clean, flat surface, with a lip for clamping material. The ends of the extension wings should be crisp and square, so they can be used for measuring.

Manufactured stands are available that are easy to set up, transport, and store; but if you're working at your home, in a couple hours, with $50 or $60 in material, you can make your own.

Measuring, cutting, and drilling

The miter saw stand shown here is made from three main parts. Only one part needs to be cut precisely. The base and top can be cut to any length and width, but the supports must be ripped to exactly the right height. If the material you're using for the top extension wings is $3/4$ in. thick, then make the supports exactly the height of your miter saw table minus $3/4$ in. If you don't have a tablesaw or can't make these rips yourself, have your local material supplier rip a piece of 1×4 or 1×6 to that width. You'll be able to cut all the pieces needed from one 8-ft. board.

Always make your first cut a practice cut, wide of the measurement mark. Once you've located the exact position of the blade on the board, creep the measurement mark slowly toward the sawblade.

For the best accuracy, try to split the pencil line in half. With your hand locked against the miter saw fence, you can position the measurement mark precisely where the blade cuts.

MITER SAW SAFETY

Power miter saws are dangerous if they're not used correctly. Make precise cuts safely on your saw by following these tips. Pay attention to additional safety tips presented throughout the book.

- **Protect your hands.** Never place your hands closer to the blade than the ends of the miter saw fence. Hold your fingers against the fence so your hand won't move, then wrap your thumb over the workpiece.

- **(Very important!) Never cross your arms.** Position your hands so that your arms are out of the way. Check your position before you cut.

- **Let the blade stop.** Always let the blade stop before lifting it out of the workpiece for safety and accuracy.

- **Protect your eyes and ears.** Miter saws are loud, so always wear ear protection. Sometimes miter saws shoot out small pieces of molding at extremely high speed, so always wear eye protection too.

Measure and cut

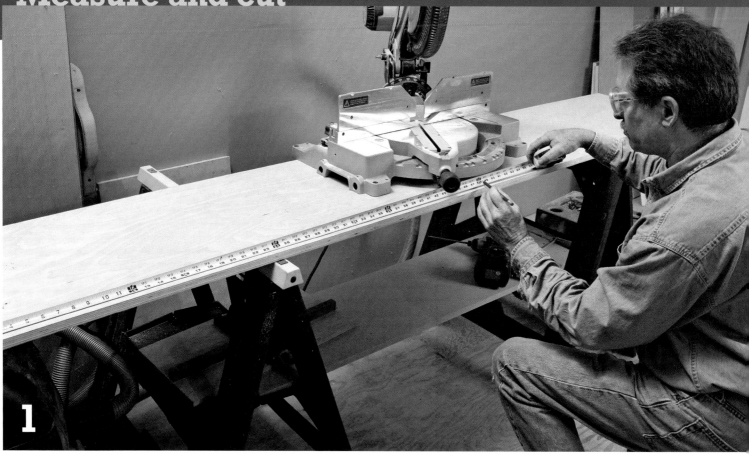

1. **Center saw on base** Set the 1×16 on top of a pair of sawhorses, then place the saw in the center. Measure from both ends to center the saw.

2. **Support workpiece with blocks** Once the stand is finished, you won't need the blocks; but for now, stack up a few blocks so the 1× support board rests flat on the miter saw.

3. **Cut support pieces** The six support pieces should be 10 in. to 12 in. long. You may have to slide the stack of blocks forward as you cut the supports.

4. **Mark repetitive stop line** After cutting the first support and before moving it from the saw, place a pencil line at the far end on the miter saw fence.

5. **Move workpiece to line and cut** After each cut, slide the board to the pencil line and make the next cut. All the supports should be exactly the same length.

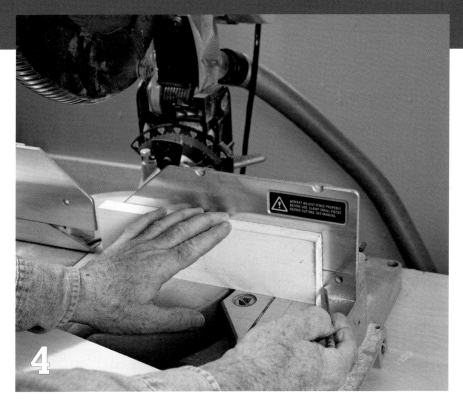

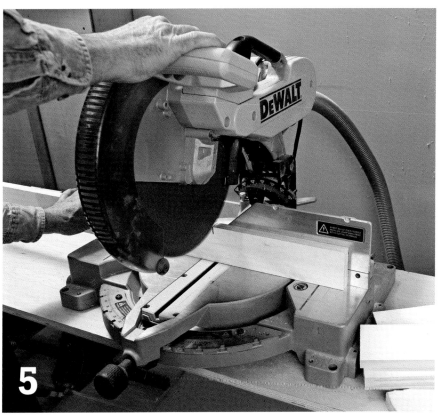

Drill pocket holes

1. **Clamp jig to worktable** Secure the pocket hole jig to a piece of 1/2-in. plywood. To support the workpiece, attach short pieces of 1× to the plywood exactly the height of the jig. Always clamp the plywood to a worktable.

2. **Adjust clamping pressure** Lock the toggle clamp down, then thread the plunger up to the workpiece. Lift the toggle clamp, and thread the plunger toward the workpiece one more turn, then back the locknut all the way to the end of the threads.

3. **Adjust stop collar** Place the bit in any of the three bushings. Slip the stop collar over the bit. Lift the bit until the tip is slightly above the jig so you won't drill into the jig. Then tighten the stop collar with an Allen wrench.

4. **Use bushings** To drill pocket holes in horizontal material, use any of the three bushings. To drill pocket holes in the ends of 3 1/2-in. material, use the outer two bushings; for 2 1/2-in. material, use the left two bushings; and for 1 1/2-in. material, use the right two bushings.

5. **Start drilling** For drilling a few pocket holes, a cordless drill works fine. For drilling a lot of pocket holes, use a corded power drill. Place one hand on the workpiece, to steady it. Hold the drill in the other hand, squeeze the trigger, wait for the bit to come up to top speed, then slowly push the bit straight down into the bushing. Feather the trigger off and on while removing the bit.

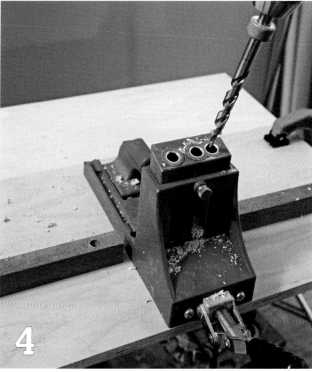

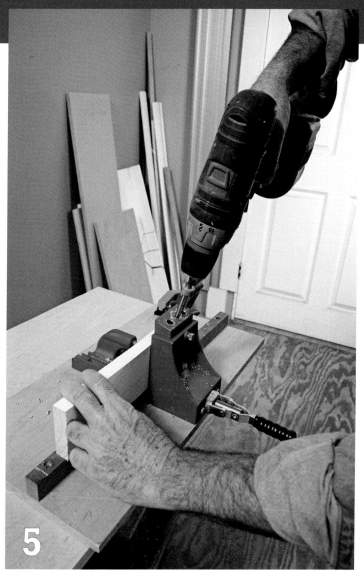

Install the supports

1. **Lay out inside supports** To allow plenty of room for your saw, measure over 2 in. from the base of the saw.

2. **Draw layout lines** Use a carpenter's square to draw layout lines for both supports, perpendicular to the edge of the base.

3. **Locate center supports** The end supports are easy, make them flush with each edge. Then measure over and mark the center of the center supports. Remember, those marks are centered on the support. Draw layout lines $3/8$ in. to one side.

4. **Glue and clamp supports** Don't attempt to drive pocket screws, or any fastener, without clamping the material in place. Otherwise, the force of the screws will push the material off the layout line.

5. **Drive in pocket screws** Set the clutch on your cordless drill to a low setting so the screw won't strip. Steady the support with one hand. Use a long square-drive bit; align the bit with the hole; and apply even but gentle force directly in line with the screw, the hole, and the driver bit.

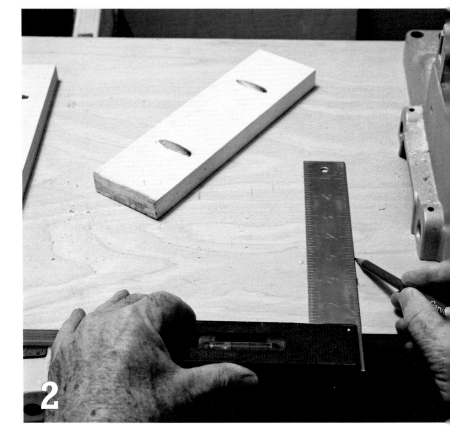

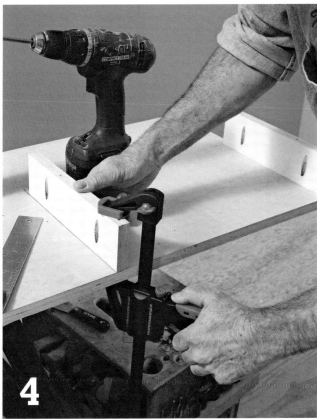

MAKE A WORKSTATION | 11

Install the top

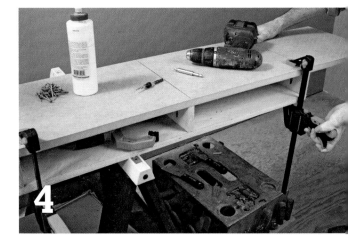

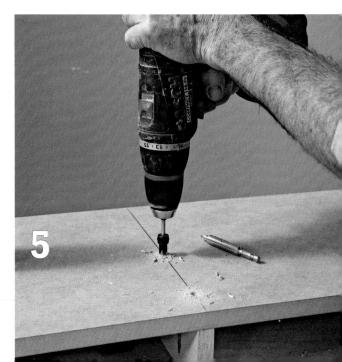

1. **Cut top in half** Some miter saws won't cut all the way through a 12-in.-wide board. No problem. Cut halfway through from one direction, then turn the board over. Align the cut with the sawblade, then cut through the other half.

2. **Position tops** Lay each top upside down behind the supports. To allow room for the saw, slide the tops 1 in. past the inside supports. The top should hang over the supports and provide a 1-in. lip for clamps.

3. **Mark centerlines** Use a square to trace centerlines on both tops for each support.

4. **Glue and clamp tops** Set each top on the supports, 1 in. past the inside support nearest the saw. Make sure the tops are flush with the base and the front edge of the supports, then clamp both tops in place.

5. **Drill countersunk holes** Use a countersink bit to drill two or three holes on each support line, 1 1/2 to 2 in. in from each edge.

6. **Fasten top** Using square-drive screws, secure the top. Drywall screws are okay for many wood-working tasks, even for fastening down the top, but square-drive screws are stronger and preferred by pros.

Make an auxiliary fence

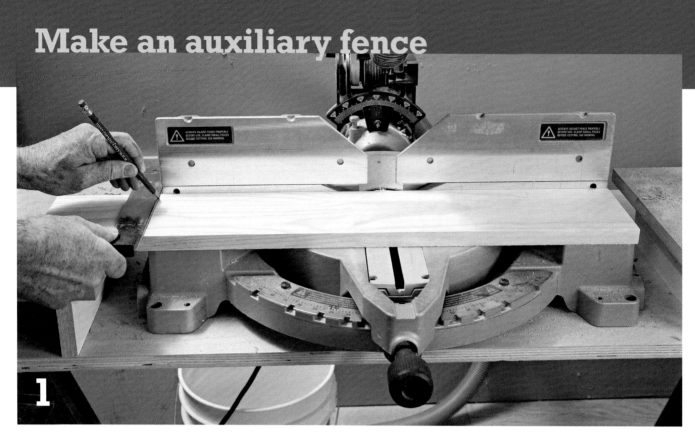

1. **Crosscut material** Rip or buy a short piece of material the same height as your miter saw fence. Cut the material the same length as your saw, measured all the way from one side of the saw to the other.

2. **Fasten auxiliary fence** The miter saw fence should have two holes in each side. Temporarily clamp then secure the auxiliary fence to the miter saw fence using four #10 by 3/4-in. screws.

3. **Cut through fence** Set the saw at 45 degrees and make a cut through the auxiliary fence. Swing the saw to the opposite 45-degree miter detent, and make a second pass through the fence. Allow the blade to stop each time.

Make a repetitive stop system

1. **Install wooden fence** Cut a 1×4 the length of the extension wing. Use pocket screws to fasten the wooden fence about 1 in. behind the miter saw fence. To prevent binding from bowed boards and moldings, do not make the repetitive stop fence flush with the miter saw fence.

2. **Make repetitive stop blocks** Cut two pieces of 1×4, each about 8 in. long., Clamp and fasten them together permanently with screws.

3. **Cut 45-degree cleat** Cut the cleat off the end of a 1×6.

4. **Drill pocket holes** Clamp the cleat securely in the pocket hole jig, with the right-angle edges down, and drill one pocket hole in each direction.

5. **Fasten cleat** Use pocket screws to secure the cleat inside the stop block. The cleat will create a perfectly square stop block.

6. **Clamp stop to fence** For repetitive cuts, measure and cut the first piece. Use the first piece to position the stop block, then clamp the block to the fence. Cut a second piece and check that it's identical to the first piece before proceeding.

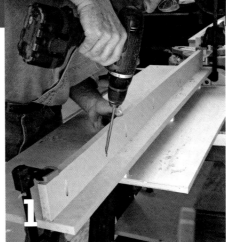

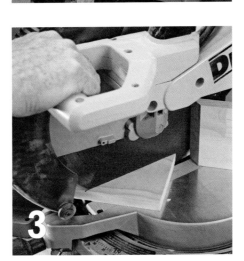
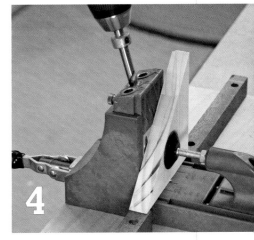

CHAPTER
2

Doors

The trim that surrounds a door frame is called *casing*, and it's always installed before baseboard and chair rail because they have to butt against it. Casing is also the easiest type of molding to install because the joinery is simple, making it the perfect first project.

I start off by explaining the details of casing joinery and describing how to measure for new casing around a door frame. I also talk about making a cut list so when you cut your casing it will be perfect the first time.

Most trim carpentry revolves around the miter saw. Here, I introduce fundamental techniques for cutting correct miters in casing—techniques that I build on in later chapters for baseboard, chair rail, and crown molding. I demonstrate a simple system for cutting casing that will help you cut each piece of molding exactly the right length every time. Later in the book, I use elements of this system to cut other types of trim.

Before installing new casing, the old molding must be removed, without damaging the wall or the jamb. I cover the best and simplest tools and techniques for that job, too. Then I demonstrate two ways for installing moldings: the time-honored one-piece-at-a-time method and a new way of preassembling casing that ensures tight miters and a neat job. By the end of this chapter, you'll be able to install new casing with confidence.

TOOL BOX

Installing casing is not only the easiest trim carpentry chore but requires the least amount of tools. Buy good quality tools. Good tools work better and last longer.

TOOLS AND MATERIALS

- Utility knife for cutting loose old casing
- Spring clamps for assembling tight and flush miters
- Trim Gauge for marking exact reveals on jambs and trim
- Wire cutters for removing old nails from a jamb or wall
- 5-in-1 tool for prying casing loose from a wall and scraping old caulking off the jamb and wall
- Prybars for removing old casing from the jamb and wall (they work best in pairs)
- Nail sets for setting nails just beneath the surface of the casing
- Thumb Saver for starting small nails in molding without hammering your fingers
- Dovetail saw/backsaw for cutting baseboard back when the new casing is wider than the old molding
- Multimaster for cutting baseboard back when you're installing casing on more than 10 new doors
- Nail gripper for driving nails without hammering your fingers or the molding
- Three 7-ft. pieces of 2½-in.-wide casing in a three-step profile pattern, made from fingerjointed, unprimed wood
- Carpenter's glue for tight, long-lasting joinery
- Nails (nail gun) for fastening the molding to the jamb and wall

Utility knife
Wire cutters
Spring clamps
Trim Gauge®
Nail sets
5-in-1 tool
Prybars
Thumb Saver®
Dovetail saw/backsaw
Multimaster®
Nail Gripper®
Three 7-ft. pieces of 2½-in.-wide casing
Carpenter's glue
Nails

Cutting casing

Avoid frustration. Always make cut lists for moldings. That's the best way to ensure a smooth, enjoyable job. With cut list in hand, you can confidently head for your saw; without one, you will likely find yourself in front of the saw trying to remember a crucial measurement, and then heading back to measure again. And with a cut list, you'll always know which way to miter your moldings—without having to close your eyes, trying to remember the room you just left.

Leg casing

At first, cutting miters in casing is confusing. To make the job easier, always place the casing with the back edge against the miter saw fence. That way, the long points of the miters will always be against the fence, and the short points of the miters—and all the measurement marks—will always be nearest to you, where you can see them best. With the measurement marks away from the fence, it's easy to guide the sawblade right to the mark, which you'll see when you cut the casing legs.

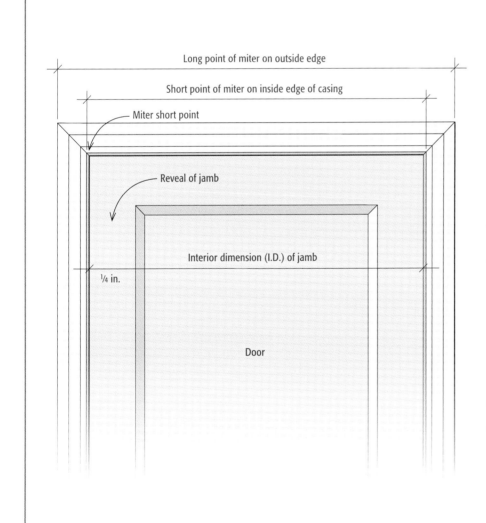

UNDERSTANDING CASING JOINERY

Before tackling any new job, make sure you see the whole picture. Casing can be confusing. A quick drawing is often the best way to visualize the job, especially when it comes to miters. Miters are angled cuts, so they always have a *short point*, where the length of the material is shortest, and a *long point*, where the length of the material is longest.

Some carpenters refer to the short point as the "heel" of a miter and the long point as the "toe," which is another way of understanding the angle of a miter. Because casing surrounds the outside of a door frame, the short points mater most: The short points of the miters are always on the jamb, on the inside edge of the casing, near the door. For casing, all measurements are taken to the short points.

Head casing

Always cut the head casings last because you can cut them from shorter pieces (sometimes from legs that you mistakenly cut too short!). The head casing is a little trickier to cut because *both* sides have miters. A simple technique makes it very easy to cut double miters at exactly the right length.

Preparation

Measuring, marking, and cutting molding take patience. Proper jamb and wall preparation takes patience too. In fact, the more care you take preparing the jamb and wall for new casing, the easier, more rewarding, and better the job will be. To speed the process, always use the right tools in the proper sequence.

CUT CASING WITH A MITER BOX

A hand-powered miter box works well for cutting small moldings, as long as it's a good one. But no matter what type of saw you use, a miter saw stand and continuous support for your material are essential.

- **Clamp your workpiece.** When using a miter box, be sure to clamp the material securely to the table and to the tool. That's the only way to ensure a perfect miter.

- **Let the saw do the cutting.** Too much pressure on the saw will distort the miter cut. Never "try" to make a miter saw cut. Allow the saw to cut by itself. Pull the blade smoothly and gently backward across the molding. Use light pressure to push the saw forward. Move your arm slowly back and forward as if it were a machine.

Clamp your workpiece.

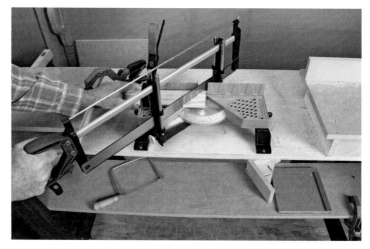
Let the saw do the cutting.

Baseboard preparation

Frequently, I install new casing without changing the baseboard. When the new casing is wider than the old casing, then the baseboard must be cut back farther from the jamb. That cut must be perfectly straight, at exactly the right distance from the jamb. If the baseboard isn't very tall, I use a handsaw; if the baseboard is big and there are a lot of doors, I use a power tool.

Installation

New adhesives, fasteners, and clamps have changed the way carpenters install casing. Often, I preassemble large casings so that I can reinforce the miters and improve the strength of the joinery. But the old method of installing casing, starting with the head piece and then following up with the two legs, is still sometimes best, especially for smaller moldings. I demonstrate both techniques here so you'll be able to work with either type of molding. No matter which technique you use, always prepare the jamb first.

Casing assembly and installation

For casing wider than 3 in., preassembly and miter reinforcement—with biscuits, splines, or pocket screws—are the best ways to ensure long-lasting miters. To improve results, use some of these techniques to guarantee tight-fitting miters around your doors.

MAKING ASSEMBLY JIGS

On some jobs, I don't have room—or enough time—to set up a full-size worktable. These simple homemade jigs make it easy to preassemble casing without a worktable, using only a miter saw stand or a sawhorse. You should have at least four assembly jigs, so several sets of casings can be put together at one time. These secure the work and keep it from slipping.

To make the assembly jigs, take a square of ½-in. or ⅜-in. plywood and cut it into roughly 8-in. squares. Then apply ¼-in. strips on two corners. If you don't have a tablesaw, a length of ½-in. door-stop molding will provide all the strips you need. Glue the strips, clamp them to the jig, and then tack them in place with brads or pins. Then add some nonstick material. Cut pieces from a router mat (www.rockler.com; $8.79), and fasten them on the back of each jig. Use a fast-acting adhesive, like 2P-10 (see Chapter 3), or contact cement to secure the nonslip material to the jig.

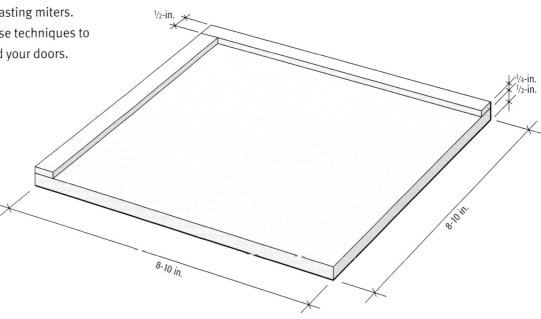

USING HAND-DRIVEN NAILS

If you're installing molding on only one or two doors, don't rush out and buy a nail gun. Driving nails by hand isn't that difficult. To protect soft wood (and thumbs and fingers) from your hammer, use the following techniques.

A. To avoid splitting the wood that you're nailing, try this old carpenter's trick: Blunt the nail tip. Nail gun fasteners have blunt tips that prevent splitting even when nailing near the edge of the casing. Hand-driven nails have sharp tips and will split the casing if driven near the edge. Blunt the tips of hand-driven nails by tapping the tip with a hammer or by cutting the tip off with wire cutters.

B. Remember to protect your fingers. Cut a narrow strip of cardboard. Poke your finish nail through the cardboard. While hammering, hold the cardboard and not the nail.

C. You might consider using plastic nail grippers. They do a good job of holding a nail firmly so it's easy to position, start, and drive the nail. Plus, a nail gripper

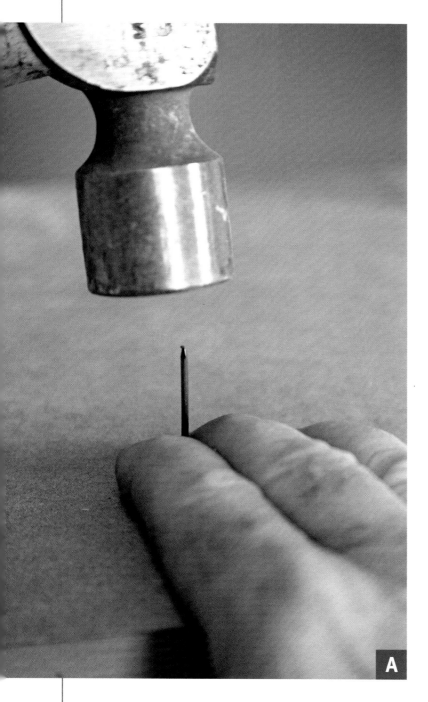

protects the casing from a missed hammer blow and stops you from driving the nail too far and striking the casing.

D. For small pins and brads, use a Thumb Saver (www.torcarr.com; $12.50/pair). This long-handled tool, with a strong magnet, secures any size nail and makes it easy to place hard-to-reach fasteners. It's very handy for assembling picture frames too.

E. Set all hand-driven nails. Drive each nail almost flush with the casing. Don't hit the casing with the hammer. Instead, stop when each nail is 1/8 in. to 1/4 in. proud (above) of the casing. Set each nail so the head is slightly beneath the surface of the casing. Use a nail set smaller than the size of the nail head.

Make a cut list

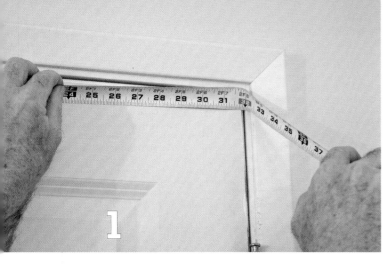

1. **Measure head casing** If the old casing isn't on the jamb, measure the inside of the jamb (inside dimension, or I.D.) and then add ½ in. for ¼-in. reveals on both sides of the jamb.

2. **Measure leg casings** Press the tape measure hook into the floor, pinch it at mid-height against the door, then stretch and curl the tape up over the top of the jamb. Measure to the inside of the jamb and add ¼ in. for a reveal.

3. **Make cut list** Write down the measurement for the head, and label it *head*. Write the measurement for the right leg, and label it *RH* write the measurement for the left leg, and label it *LH*.

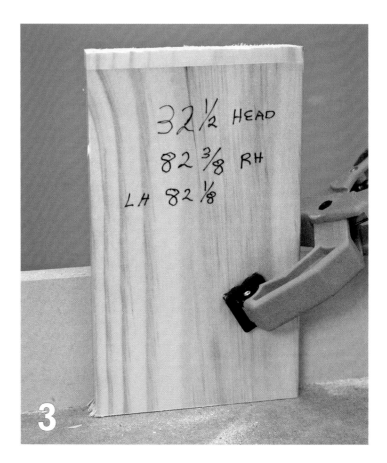

Cut back the baseboard

1. **Mark baseboard** Use a short piece of casing to trace a line on the baseboard at exactly the right location. Be sure to allow for a reveal on the jamb—hold the casing ¼ in. back from the edge of the jamb.

2. **Make cut** Using a dovetail saw or backsaw, guide the saw near the floor with one hand. Slide the saw gently up and down with your other hand. Don't push on the blade or try to cut fast. Constant, even, and light pressure is the secret to a clean cut.

3. **Use power tool** If your project involves more than 10 doors, the Fein Multimaster™ is the ideal tool for cutting back baseboard. This power tool is very easy to control and cuts with little noise, dust, or vibration. The tool and the blades are expensive, but the Multimaster is handy for a variety of difficult chores, such as scraping adhesive; chipping out tiles; cleaning grout; and cutting metal, concrete, and drywall.

Cut the legs

1. **Cut legs first** Place one piece of casing on the left side of the saw. That will be the left-hand leg. Place another piece of casing on the right side of the saw, that will be the right-hand leg.

2. **Measure up from bottom** Hook your tape measure on the bottom of the casing, stretch the tape up toward the saw, and make a measurement mark on the font edge of the casing. Do the same for the opposite leg.

3. **Cut right leg** Swing the saw toward the right to 45 degrees. Place your right hand at the end of the miter saw fence, wrapping your thumb around the casing. Position the measurement mark about 1 in. from the blade, and make a practice cut.

4. **Look from front of saw** Sighting down the sawblade is the hardest way to align the blade with the measurement mark. The measurement mark is easier to see from the front of the saw, even with the blade spinning. With your thumb wrapped around the casing, slide the casing toward the blade, creeping the measurement mark forward until the blade cuts right on the mark.

5. **Cut left leg** To cut the left leg, swing the saw to the left. Place your left hand at the end of the miter saw fence and wrap your thumb around the casing. Make a practice cut, wide of the measurement mark, then creep the measurement mark up to the blade.

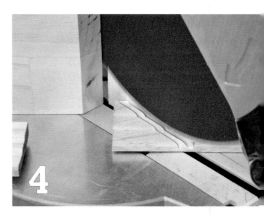

Cut the head

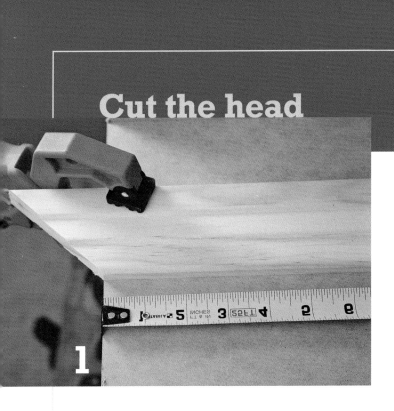

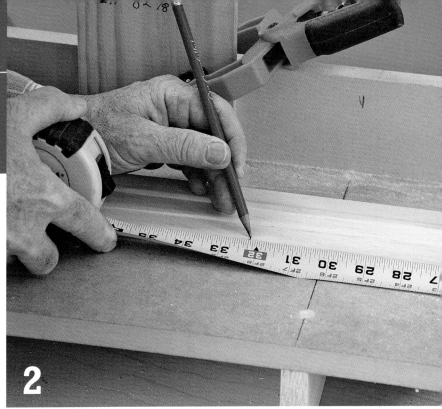

1. **Miter right end** Clamp the casing to the end to the miter saw extension table, with the short point flush with the edge. Hook your tape measure on the edge of the miter-saw table, then pull the measurement.

2. **Measure with sharp pencil** Use a no. 2½ pencil to make a crisp, fine line. Never round off fractions. Always make measurements exact, to within $1/32$ in., which is a little more or a little less than the nearest $1/16$ in. mark on the tape measure.

3. **Make practice cut** Hold the casing with either your right or your left hand placed at the end of the miter saw fence. Wrap your thumb over the front edge of the casing. Make a practice cut about ½ in. away from the measurement mark.

4. **Creep up on mark** The more you use a miter saw, the closer you'll make your practice cuts and the fewer practice cuts you'll make. But don't rush the learning process. Cutting right on the measurement mark is what matters most.

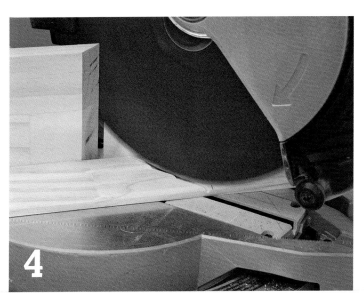

Prepare the jamb

1. **Cut caulk joint** Use a sharp utility knife, angled between the wall and the casing, to cut through old caulking and paint. That's the first step in breaking the casing loose.

2. **Work 5-in-1 tool** A 5-in-1 tool is a hybrid scraper/prybar/can opener and is a must for removing moldings. Rock the sharp, stiff blade back and forth, working it under the casing.

3. **Wiggle in small prybar** Use the 5-in-1 tool to lift the molding away from the wall just enough to wiggle in the small prybar, then work the tools in opposite directions—use one to pry against the molding; use the other to pry against the wall.

4. **Finish with medium prybar** Once the gap is large enough, slide the short end of the medium prybar as far under the casing as possible. It's best if the prybar pries against the wall under the casing, rather then behind it, where marks might show. Work your way down the jamb with the medium prybar, removing the casing.

5. **Clean wall** Cut or remove all nails with wire cutters or pliers, then scrape clean any caulking or paint buildup using the 5-in-1 tool.

Install mitered casing

1. **Mark reveals** Before installing any casing, draw reveal lines on the jamb ¼ in. back from the inside edge. A pair of scribes will do the job but a marking gauge speeds up the task. The adjustable Trim Gauge can also be used for a variety of reveals or back-set layouts.

2. **Align miters** Line up the miters with the reveal marks. Tack the head casing to the jamb. If you're using a nail gun, shoot one 23-ga. or 18-ga. brad near the center of the head casing.

3. **Apply glue to miters** Spread a thin layer of carpenter's glue on both miters before assembling the casing.

4. **Tack leg casing to jamb** Position the miter so the molding profiles align. Place the first nail about 4 in. below the miter. Drive a second nail about 8 in. below the first nail.

5. **Attach spring clamps** A glue joint will not be strong unless it dries under pressure. Before driving more fasteners, install spring clamps on both miters. Adjust the clamps and the miters so the profiles are aligned and flush.

6. **Nail off casing** Finish fastening the casing by driving brads or pin nails every 8 in. to 12 in. through the casing into the jamb. Fasten the casing to the wall every 12 in. to 14 in. To reach through the casing and drywall and penetrate the studs by least ¾ in., use 15-ga. 2-in. nails.

30 | TRIM MADE SIMPLE

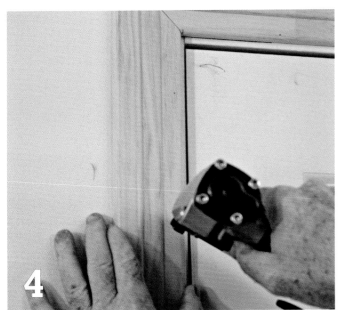

Preassemble casing

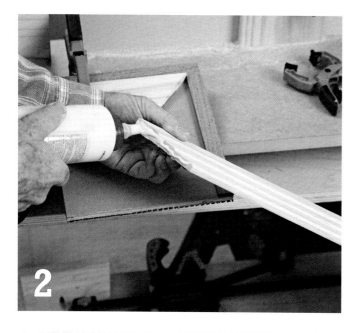

1. **Clamp head to jig** Use A-clamps to secure an assembly jig (see "Making Assembly Jigs" on p. 21) to each end of the head casing. Make sure the casing is tight against the stops.

2. **Glue miters** Spread a thin layer of carpenter's glue on each miter.

3. **Clamp legs to jig** Rest the head casing and assembly jigs on the edge of a sawhorse or on your miter saw stand. Tilt the head casing and each leg into position, squeezing the miters tightly closed. Use A-clamps to secure the legs.

4. **Clamp miters** Glue joints won't be strong unless they dry under pressure, and putting glue under pressure helps it set faster. Use a wrench to spread each spring clamp as wide as possible. Position the clamps on top of each miter.

5. **Carry frame to wall** Strong A-clamps make it possible to move the casing off your work area and store it temporarily against a wall while the glue sets. Wait 10 minutes or 15 minutes before installing the frame as described in Step 6 on p. 37.

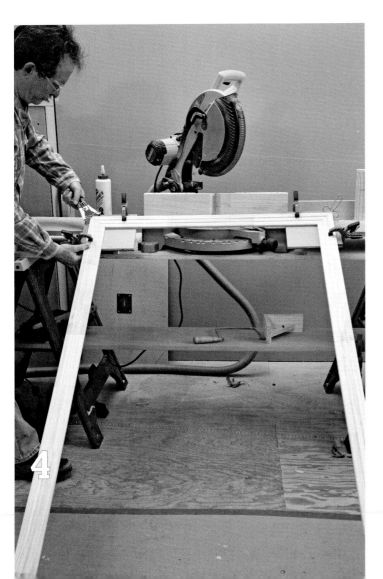

CHAPTER 3

Windows

Window and door casing are meant to mimic classical architecture, to resemble columns and pilasters on Greek and Roman buildings. Window casing can be installed several different ways, but one of the most attractive styles is using a stool and apron at the bottom of the frame, along the windowsill. Traditional stool and apron originate in classical architecture too, from the molding that caps a tall plinth or column base.

In this chapter, I demonstrate a simple foolproof method for installing stool molding, including how to scribe and cut the stool so that it fits snugly against the window frame. Next, I explain how to cut the apron and the delicate self-return caps. Finally, I repeat some of the techniques introduced in Chapter 2 plus present a few new ones, so you can complete the casing.

Picture framing a window with casing is boring. Install a stool and apron to dress up your window.

Cutting and installing the stool

Even though it's only one piece of molding, installing window stool takes more time than casing a door—and more tools, too. These easy-to-follow, step-by-step directions will guide you through the process of laying out, cutting, making the supports, scribing, and installing the window stool. Don't cut any corners.

Careful layout

Installing the stool correctly depends on careful layout. Try not to use a tape measure. Instead, draw all the details in full scale right on the window and the wall before cutting a single piece of wood. Take your time. The difference between an old carpenter and a young carpenter is patience, that's why older carpenters make fewer mistakes.

TOOL BOX

Like the tools listed in preceding chapters, most of the tools needed for installing window casing can be used for other tasks; each one forms an important part of a finish carpenter's toolkit.

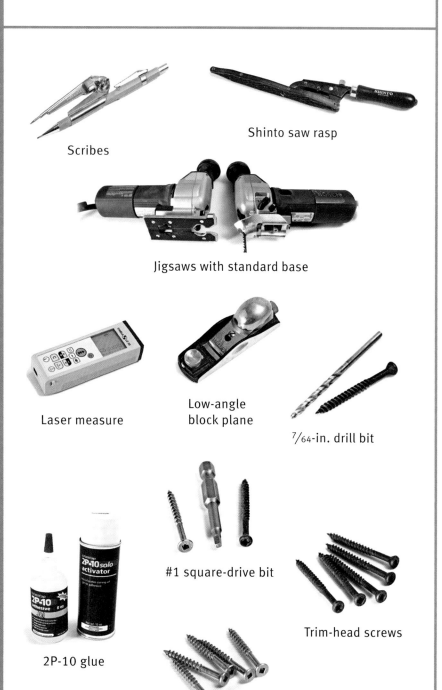

Scribes

Shinto saw rasp

Jigsaws with standard base

Laser measure

Low-angle block plane

7/64-in. drill bit

2P-10 glue

#1 square-drive bit

Trim-head screws

#6 square-drive screws

TOOLS AND MATERIALS

- Scribes for scribing molding to fit tightly against a wall, jamb, or other surface

- Shinto saw rasp for rough filing or fine adjustments on notches around corners cuts

- Jigsaw for cutting to scribed lines quickly and enjoyably

- Low-angle block plane for following a scribed line and shaving small amounts of wood from molding

- Laser measure for measuring over your head or for measuring long molding

- 7/64-in. drill bit for drilling trim-head screw pilot holes

- #1 square-drive bit for driving screws

- 2P-10 glue for fastening small, delicate pieces of moldings

- Trim-head screws for fastening stool to casing

- #6 square-drive screws for fastening stool to windowsill

- Four 6-ft. pieces for casing

- One 6-ft. piece for stool

End cuts

Stool is available with different edge profiles: bullnose, square, and beveled, like the style in this example. Beveled stool adds a little drama to the molding, yet it's easy to work with. The end cuts can be made on the miter saw, without any other special cutting tools.

Stool supports

Scribing stool to fit tightly against a windowsill and wall—like scribing any molding—is a delicate operation. Achieve nearly perfect success on the first try by securing the workpiece to temporary supports before attempting the scribe.

Prepare to scribe

Scribing is a technique that carpenters have used for centuries, and it's one skill that every carpenter should become proficient at. There's no better way to fit two pieces of material perfectly tight, no matter how uneven the original joint. Take your time setting up the scribe, and the results will be pleasing.

Scribing

Using a pair of scribes is easy. Just remember: Always keep the scribes perpendicular to the surface you're scribing from. You can rotate, lift, or lay down the scribes, but you can't *tilt* them. Tilting the scribes changes the distance between the scribe line and the material you're scribing from.

Cutting to the scribe line

Cutting to a scribe line must be accomplished with as much care as making the scribe itself. For that reason, and because it's the only safe way to work with tools, never use your hands as clamps. Never use any hand-held tool—whether it's a power tool or not, without clamping the workpiece securely to a worktable. Guiding and controlling a tool requires two hands and total concentration on the cut.

AN ANATOMY OF WINDOW TRIM

Like many carpentry tasks, trimming out a window always begins at the bottom, with the stool molding, which forms the foundation for all the other trim on the window. Get the stool installed correctly, and the rest of the job is easy. Trimming windows is no different from installing casing around a door.

The width of stool depends on several things: how far the windowsill is from the face of the wall, the thickness of the casing, and the design of the apron. A 2-in. stool works well for many applications. If you can't find stool the right width and you don't have a tablesaw, ask your home center to rip the material for you.

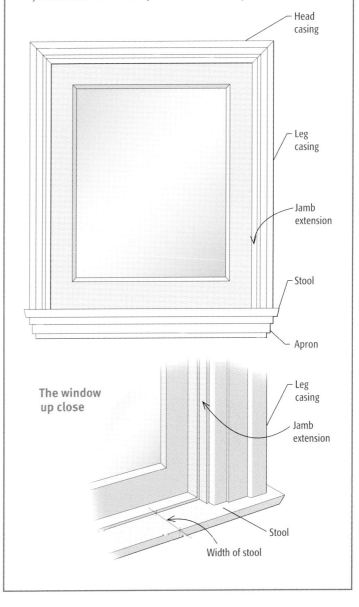

Pocket holes

There are several ways to attach the stool to the windowsill. Splines, biscuits, and dowels are often used, and they work well but require additional tools. This pocket screw method is easy to do with the tools you should already have. Unfortunately, because stool is often narrow, you can't use a pocket hole jig and the pocket holes must be drilled by hand.

Stool installation

Normally, when using pocket screws the workpiece is secured by clamps. But clamping the stool to the windowsill isn't always possible. In this case, rather than using clamps, I demonstrate how to use pocket screws to help position the stool at exactly the right location.

Cutting and installing the casing

Now that the hard part is finished, and we have a solid, level foundation for the casing, it's time to trim the window. Measure each piece carefully, especially the legs. They should fit tight against the stool.

Measuring casing

In Chapter 3, we covered measuring, cutting, and installing door casing, so you might wonder why we're covering casing again. Repetition is the secret to learning anything, especially carpentry. Besides, with finish work, there are always new techniques to learn. For this project, we measure both with a laser "measure" and with a tape measure. Stretching a tape across a 6-ft. opening isn't too difficult, but using a tape to measure a 12-ft. opening accurately is impossible to do alone without driving temporary nails to secure the tape. Fortunately, high-quality lasers are now accurate enough to use for measuring finish work.

Cutting casing

I described how to cut casing in Chapter 2. But with trim carpentry, there's more than one way to do almost everything. Experiment with different methods of cutting to find the one that suits you best.

Casing installation

Preassembling casing is faster and the results are more satisfying than assembling the pieces individually on the wall. But all measurements must be precise, and all pieces must be cut accurately. If you're not confident that your pieces are cut perfectly, test-fit them before assembling the miters.

Cutting apron

The apron is made from the same material as the casing, but cutting apron is completely different from cutting casing. Casing is cut lying flat on the base of the of the saw—think of the base of the saw as the wall. But apron is cut standing up, flat against the miter saw fence. Think of the fence as the wall.

Return cap assembly

As I said earlier, self-return caps are small and delicate. Securing these pieces of molding with nails isn't always possible. Unless the molding is exceptionally large, even a brad nail will split a cap. Years ago, we used to glue and tape self-return caps, but now we use fast-acting glue to make perfect joints quickly.

Apron installation

Fastening the apron to the wall is the last step; and though the self-return caps are preassembled, you can't just nail the apron to the wall. The piece must be secured tightly against the stool, both so the joint is tight and so the apron helps support the stool.

Lay out the stool

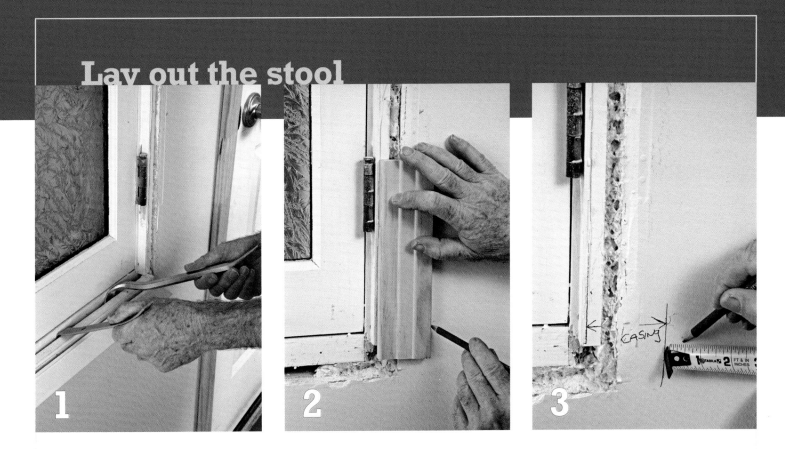

1. **Remove casing** Remove all the casing from around the window, but leave the jamb extensions if they're in good condition. Because the stool will replace the jamb extension along the windowsill, remove the casing and the jamb extension at the bottom of the window.

2. **Lay out casing and jamb reveal** Use a small piece of casing as a full-scale model. Hold the casing against the jamb and trace a pencil line along the back of the casing.

3. **Add for stool reveal** The stool should extend along the wall past the casing, creating a stool reveal. A reveal of ½ in. is nice for a stool that's cut square on the edges. For beveled stool, make the reveal ¾ in.

4. **Measure stool length** Lay out the same lines for the casing and stool at the other end of the window. Drive a finish nail into the drywall at the mark made at one end of the stool. Hook a tape measure on that nail and measure across the window to the opposite end of the stool.

Cut the stool

1. **Find bevel angle** Cut a short piece of stool, stand it on edge near the opening of the throat guard, then swing the miter saw until the edge of the throat guard and the molding are parallel.

2. **Cut one end** Cut the first end at the bevel angle. For most stools, that angle is 15 degrees.

3. **Cut other end** Hook a tape measure on the long point of the bevel; measure; and then make the second cut, swinging the saw in the opposite direction. Always hold the stool with the bottom flat against the miter saw fence. The short point of the bevels should always be against the fence.

Make stool supports

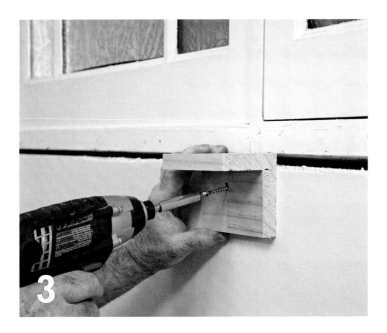

1. **Make support blocks** Cut four scrap pieces of stool or 1×4 about 4 in. long. Form a L-shaped support and fasten the two pieces on edge with brads.

2. **Reinforce supports** Predrill a countersunk pilot hole in each support, and drive in a 1¾-in. drywall screw. *Note*: A pilot hole is approximately the size of the screw shank, so the screw threads will still bite into the wood.

3. **Secure supports to wall** Predrill a through hole, and drive a long screw through the support into the wall. *Note*: A through hole is slightly larger than the screw threads, so the screw will not bite into the wood. Keep the top of the support flush with the bottom of the windowsill.

Set up the scribe

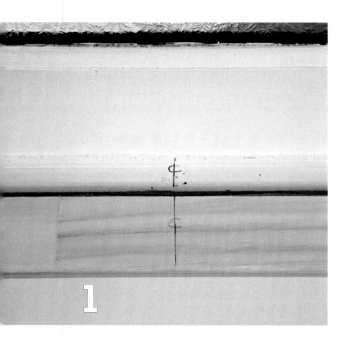

1. **Center stool** Mark a centerline on the stool and the windowsill. Set the stool on the support blocks and line up the center marks.

2. **Clamp stool** Secure the stool to the support blocks, checking that it stays centered on the windowsill.

3. **Make stool parallel to window** If you start with the stool parallel to the windowsill before scribing the notches, the stool will be parallel after cutting the notches. Measure the distance between the stool and the windowsill at one end.

4. **Measure opposite end** Check the opposite end. If necessary, move one end a little so both measurements are the same. Check that the clamps are secure.

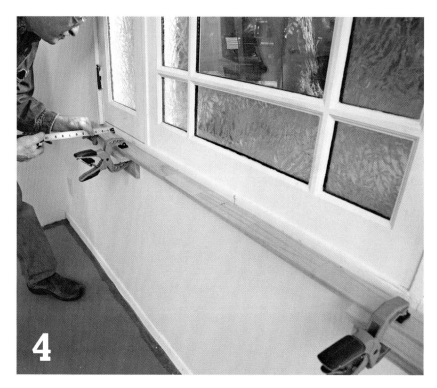

Scribe the stool

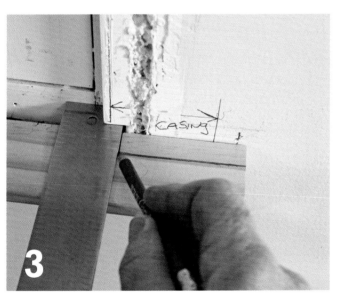

1. **Spread scribes from sill to stool** Spread the scribes so that the pencil lead just reaches the stool. That's exactly how much material must be removed from the horns for the stool to reach the windowsill.

2. **Scribe wall** Keep the scribes perpendicular to the wall. You can lift them up or lay them down, but the point of the scribes and the pencil lead must always be perpendicular to the wall.

3. **Scribe shoulder cuts** Use a carpenter's square to scribe the shoulder cuts. Those lines don't have to be perfect because the jamb extension hides most of that joint.

4. **Scribe both sides** Move to the opposite end of the stool, and scribe that side exactly the same way. Don't change the spread on the scribes! If the pencil lead doesn't reach the stool or if it's too far past the stool, start over.

Cut the stool

1. **Use fingers to guide saw** Hold the jigsaw firmly from the top handle. Pinch the leading edge of the saw base with your free hand and guide the saw along the scribe line. Holding the saw base with your fingers will ensure that you always know where those fingers are. *Never* wrap your fingers underneath the workpiece.

2. **Clean up with rasp** Inexpensive four-way wood files never work well. The serrations never reach all the way to the edge of the file. A Shinto saw rasp has two cutting edges, one side is coarse; the other is fine. It's the best tool for cleaning up a notch, right to the center of the scribe line, even in a tight corner.

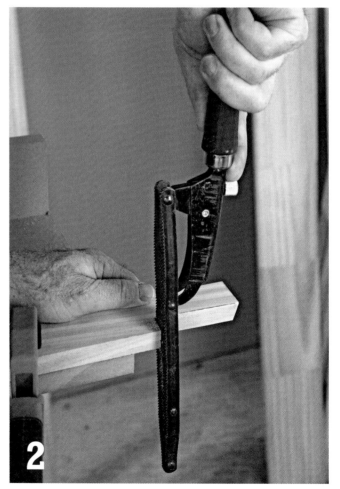

3. **Check fit** Scribing is almost always a two-step process. The first step must be made very carefully so that the second step is easy and quick. Notice the small gap near the left hand side. That's because there's a small bow in the middle of the windowsill.

4. **Plane center** Most block planes don't cut easily because of the angle between the blade and the bottom or sole of the plane. A low-angle block plane cuts easily and glass smooth—if it's kept sharp. If you can't sharpen your tools yet, ask a local tool supplier for the name of a good sharpening company near you.

5. **Recheck fit** A few strokes with a sharp plane or one or two more passes with a Shinto saw rasp, is usually all it takes to perfect a scribe cut. The horn should fit snugly against the wall, just as the stool touches the windowsill.

Install the stool

1. **Mark reveal line on sill** Use the Trim Gauge or a set of scribes to mark a reveal line on the edge of the windowsill. The reveal on the sill should match the reveal on the door jamb and the reveal you'll be using on the window jamb.

2. **Slide Trim Gauge** Rather than making a series of short marks, hold your pencil firmly against the underside of the Trim Gauge, then slide the tool along the window frame, leaving a straight clean line on the windowsill.

3. **Start at ends** Start fastening the stool at the ends. The jamb extensions will help stop the stool from creeping up on the sill while you're tightening the fasteners. Notice how much of the scribe line is visible after you've driven the screws at both ends.

4. **Secure center** As the screws tighten, the stool will creep up the windowsill. Plan for that movement by pulling the stool down, approximately 1/8 in. beneath the reveal line, before driving the center screws.

Measure the casing

1. **Mark reveal lines** Use the long side of the Trim Gauge. The offset built into the gauge is ¼ in. Slide the tool along the jamb, holding the pencil firmly against the gauge.

2. **Measure head** To stretch a tape measure across a wide window or door opening, hold the tip of the tape 2 ft. to 3 ft. back from the end. Butt the tape measure against the jamb or jamb extension on the far side, then read the measurement at the reveal line on the near side. Add ¼ in. for the reveal on opposite jamb.

3. **Measure with laser** Some lasers are accurate to ¹⁄₁₆ in. over 300 ft. This tool will read a measurement from the front or the back of the housing. Press the large green button once to turn the laser on. Align the red dot on the opposite jamb, then press the same button once more. And you might not need to make a cut list; the laser remembers the last four measurements you took.

4. **Measure with tape** Tape measures will always be a carpenter's most frequently used tool. Learn how to use one. With your free hand, pinch the tape against the jamb while pressing the tip against the stool. Hold the case in your opposite hand and bend the tape against the wall. Continue to feed tape up the wall. Allow the bend to roll past the reveal line. If you need more tape to reach the top of a tall window, pinch both ends against the wall and pull more tape out of the dispenser case.

Cut the casing

1. **Cut head** For short pieces of head casing, cut the right-hand miter, then flush the short point with the edge of the throat guard, and hook your tape measure on the guard.

2. **Use repetitive stop** If both the left and the right legs are the same length, cut the miters first, then use a repetitive stop to cut the legs.

3. **Reverse molding** For the left-hand leg, the long point will be against the fence. For the right-hand leg, the short point will be against the fence.

4. **Preassemble miters** Window casing, even for windows as wide as 10 ft., is easy to preassemble at a miter saw station using preassembly jigs and miter clamps.

Install the casing

1. **Tack legs** Hold one leg on the reveal line and drive a single brad about 12 in. below the miter. Then check that the opposite miter is also on the reveal line. If both pieces aren't precisely on their lines, split the difference.

2. **Nail off casing** Secure the casing to the wall with 15-ga. 2-in. or 2½-in. nails. Drive nails every 14 in. to 16-in. around the perimeter of the casing.

3. **Predrill for screws** The head on a trim-head screw is slightly bigger than a 16d finish nail, about 3/16 in. Drill a 3/32-in. hole through the bottom of the stool, but stop before drilling into the bottom of the casing.

4. **Screw stool to casing** Slowly drive a trim-head screw through the stool and up into the casing. Don't rely on the screw to pull the stool tightly against the casing. Push on the casing with your free hand. Otherwise the screw might strip in the casing.

Cut the apron

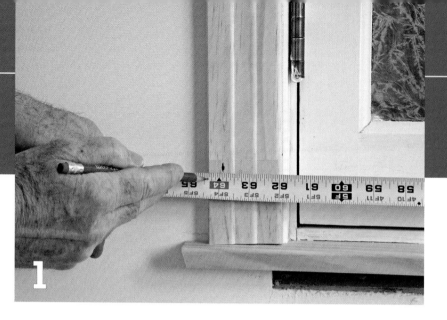

1. **Measure apron** The apron should align vertically with the outside of the casing. Hook a tape measure on the outside of the casing, and measure to the outside of the opposite casing.

2. **Cut caps first** To make sure you cut the self-return caps correctly and to be sure the grain and profile fit perfectly, cut all caps as close as possible to the miters. That means cutting the right-hand cap first, before cutting the right-hand miter in the apron.

3. **Let saw stop** For both safety and craftsmanship, never lift your saw before the blade stops. Self-return caps are very delicate pieces of molding. Air movement from the sawblade can draw a small cap right into the spinning teeth, ruining the molding and firing a projectile from the saw.

4. **Cut cap, then miter** Before cutting the miter, cut the molding off square, about 2 in. beyond the measurement mark, cut the cap, then cut the miter.

5. **Measure from long point** Swing the saw, and cut the miter on the right-hand end (keep the short point against the fence). Then hook a tape measure on the long point, across the face of the molding, and measure the length of the molding.

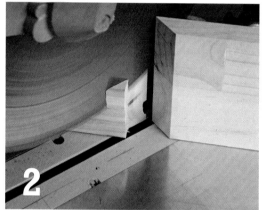
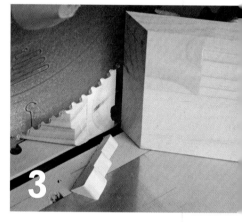

Assemble the self-returns

1. **Spread glue on smaller piece** Fast-acting 2P-10 glue is similar to Super Glue™ but stronger (www.fastcap.com). Do your best to keep your fingers away from the glue and especially from the aerosol activator, which sets off the glue almost instantly.

2. **Hinge miter closed** The *2P-10* stands for "two-part glue that dries in 10 seconds." Get your ducks in a row. Align the long points of the miter, then hinge the miter closed slowly.

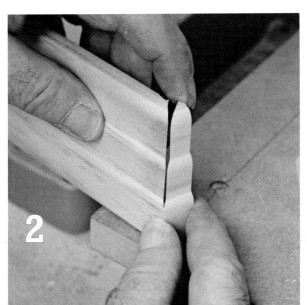

Install the apron

1. **Secure apron with clamps** In this case, the apron had a small bow, so I needed only one clamp, at the center of the stool, to squeeze the two moldings tightly together. Always nail the apron to the wall first, using 2-in. nails.

2. **Fasten stool** Use 2-in. nails to fasten the stool to the apron. Drive the nails from the top of the stool down into the apron. Be sure to clamp the two moldings securing at every nail, especially if you're using hand-driven nails; otherwise, the force of the nails might cause the two pieces to separate a bit.

CHAPTER

4

Baseboard

The joinery in baseboard forms the foundation for nearly all the joinery in finish carpentry, which makes perfect sense because baseboard is meant to replicate the foundation—the plinth—of a classical column. Though casing is the first molding profile noticed in a home, baseboard is often the first molding that an apprentice carpenter learns to install, and for good reason. The first big challenge in finish carpentry is learning how to identify and cut inside and outside corners—both miters and copes. Carefully follow the lessons in this chapter, practice on scrap before making your actual cuts.

Planning baseboard joinery

Most carpenters are never taught how to recognize inside and outside corner miters. Instead, apprentice carpenters are assigned a closet in the back of a home and told to figure it out on their own. That's a tough way to learn and explains why many carpenters never master the simple basics of miter joints. Watch most carpenters at work, and sooner or later you'll see them close their eyes and try to visualize the direction they need to miter a piece of molding. Learn these basic rules—the short-point/long-point method—and you'll never wonder which way to miter your moldings, even when you're cutting them upside down and backward (see Chapter 6).

Make a cut list

Most homes and rooms are rectangles; most joinery is inside corners and butt joints. It's all pretty simple. But remembering measurements and corner joints is tough. And walking back and forth from your saw to the room is time-consuming. Always make a cut list. For right-handed carpenters, walk into a room, pick a wall and move to

The short-point/long-point method

For outside corners, the short point of the miter is always at back of the molding, against the wall and against the miter saw fence. For inside corners, the long point of the miter is always at the back of the molding, against the wall and against the miter saw fence. For termination against casing or other moldings, pilasters, or cabinets, baseboard is cut with a butt joint.

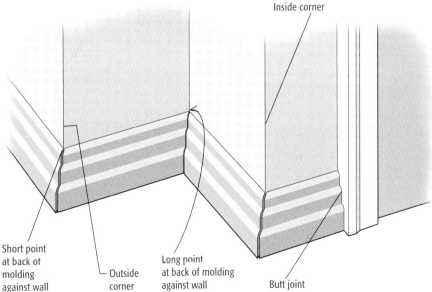

Inside corner

Short point at back of molding against wall

Outside corner

Long point at back of molding against wall

Butt joint

TOOL BOX

By now, you should have a fair collection of tools. The tools introduced in this chapter will fill out your toolkit and enable you to tackle the chapters on chair rail and crown molding. In fact, you'll be tooled up and prepared for almost any trim carpentry project.

TOOLS AND MATERIALS

- Protractor for measuring miter angles for corners

- Coping saw for cutting coped joints in baseboard, chair rail, and crown molding

- Coping foot for making cope cuts in molding easily with a jigsaw

- Stud finder with powerful magnets on the back for locating wall studs so nails in molding penetrate into solid framing

- 3½-in., three-step baseboard, enough footage for the job plus extra for practice cuts (and so you won't be nervous about making mistakes; count on making mistakes)

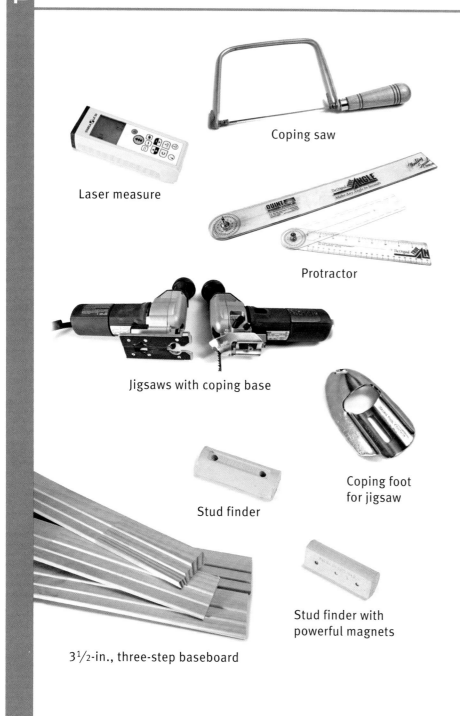

Laser measure

Coping saw

Protractor

Jigsaws with coping base

Coping foot for jigsaw

Stud finder

Stud finder with powerful magnets

3½-in., three-step baseboard

MEASURE LONG RUNS CAREFULLY

Measuring long runs is difficult, and the learning curve is steep—at least it used to be. Here are a few techniques that will make the job easier for new carpenters and go more quickly for seasoned carpenters.

- **Read a bent tape measure.** Most carpenters learn the hard way how to read a bent tape measure—by trial and error. But it's never exact and always requires a little guesswork.
- **Use a block.** Avoid bending a tape measure by cutting a block of baseboard exactly 10 in. long. Measure to that block and then add 10 in.
- **Use a laser measure.** Laser distance-measuring tools are becoming more and more popular. The price of these tools is dropping steadily, and they're now extremely accurate, making them ideal for measuring inside corner pieces, especially on long walls.

your left. Carpenters who use a coping saw in their left hand, might find it easier to move toward their right. You see why when we get to coping.

Outside corners

Outside corners must be mitered. Because of that, the joints must be cut precisely to fit the wall. Measure these pieces carefully, and expect to spend some extra fiddling time at each corner.

Cutting baseboard

With the miter saw set up as shown in Chapter 1, you'll be able to cut baseboard precisely and easily. In fact, it should be fun. If it's not fun, then you're doing something wrong. You should always have fun doing finish work. Having a cut list makes the job even easier. With your molding standing up against the saw fence, the right end of the molding is the right-hand corner, and the left end of the molding is the left-hand corner.

Cutting the cope joint

On short pieces, I always cut the cope joint before cutting the other end of the piece off at the measurement mark. That way I have enough material to clamp securely to my work station. On long pieces, I cut the piece to the measurement mark then make the cope cut. (To learn how I cope a joint, see Chapter 6, pp. 97–98.)

Inside corners

Inside corners, if they're coped, are pretty easy to cut if you work carefully and patiently. Cutting outside corners requires the same care and patience but uses a slightly different set of techniques.

Installation

Before you start kneeling on the floor, buy a good pair of knee pads or place a piece of carpet or foam beneath your knees.

Installing baseboard is a critical part of trim carpentry. The techniques used vary from wall to wall. You'll find yourself coming back to some of these same techniques when you're doing other carpentry tasks. With the cope aligned, tap the back of the baseboard into position against the wall. A well-measured piece should be snug but not move the casing. The cope joint should close up almost watertight.

UNDERSTANDING PROTRACTORS AND MITER SAW ANGLES

Don't waste your time bisecting angles with a compass or cutting little blocks of wood to determine a miter angle. Use a protractor. I tried using a protractor about once every year for 20 years before I finally learned that the angles on most miter saws don't work with a protractor. Use an indelible marker (like a Sharpie®) to write the protractor angles on your miter saw. Yes, you have to push the miter saw *past* 45 degrees to reach 43 degrees—it's a sharper angle.

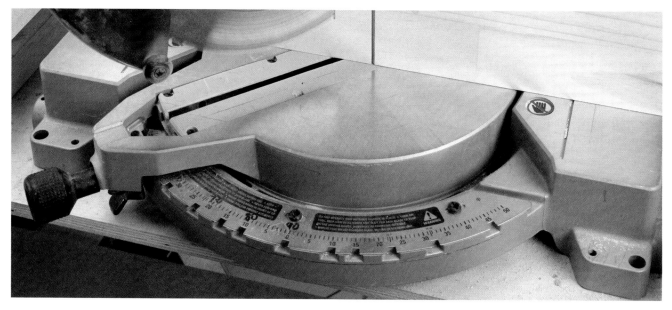

Long walls

Not all of finish carpentry can be taught through simple rules. Some of it must be learned through experience. Installing molding on long walls is one example. To get tight-fitting joints on long walls, carpenters frequently "pressure-fit" the pieces: They cut them a little long so the molding snaps into place and the joints end up perfectly tight. But learning just how much pressure to apply takes experience. Experiment, and you'll find the right fit.

Fitting outside corners

Miters on outside corners are the first things people see in molding. Fitting them requires a unique set of techniques and an eye for problem solving. When you reach an outside corner, down shift into slow mode. If you make a mistake, don't waste it. Rather than getting angry with yourself and losing patience (and making more mistakes), say these magic words: "How'd I do that?" Understand why you made the mistake so you won't make it again. For instance, you may look back at your cut list and see that you wrote down the wrong measurement mark. We all make mistakes. Rechecking your measurements is always a good idea.

Nailing baseboard

I fit baseboard before I fasten it. That way I'm not carrying a nail gun and hose with me while I'm fiddling with the drywall, notching inside corners, and generally preparing the molding.

JIGSAW SAFETY

Follow these two safety tips exactly, so you'll know precisely where your fingers are all the time—otherwise, you might be looking around the floor for one of them. As you can see in the photos, I always have two hands on the saw, with one of them holding the molding,

- **The push position.** When pushing the saw into the molding, hold the saw upside down in your right hand with the blade pointing away from you. Keep your left thumb against the guard on the coping foot and curl your fingers into a relaxed fist—so they'll never get near the blade.

- **The pull position.** When cutting toward you, hold the saw in your right hand with the blade near your wrist. Touch the motor of the saw with your left fingers. On my saw, I like to hook my fingertips on the orbital adjustment lever. Find a comfortable position on your saw for your fingers and always put them in the same place, so they'll never get near the blade.

Make a cut list

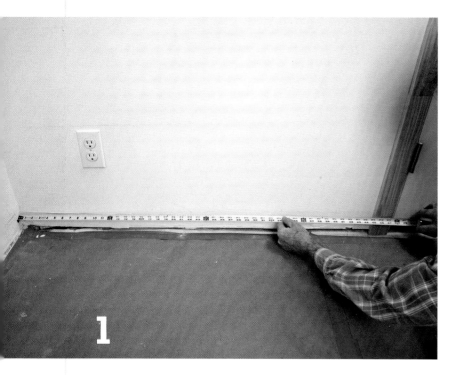

1. **Measure pieces under 6 ft. precisely** The first piece for this project measures 45 1/8 in. and has an inside corner on the left and casing on the right.

2. **Mark cut list** Write the measurement in the center of the cut list, and write the letter B on both sides. The right-hand end butts the casing, so it must be a butt cut. The left-hand end is butt cut, too, because we're coping all inside corners and the first piece into a coped corner is a butt cut.

3. **Measure short pieces a hair short** Sometimes short pieces are hard to install. I measure them a hair short—about 1/32 in.—especially when they butt against casing. If you measure them too long, you might move the casing.

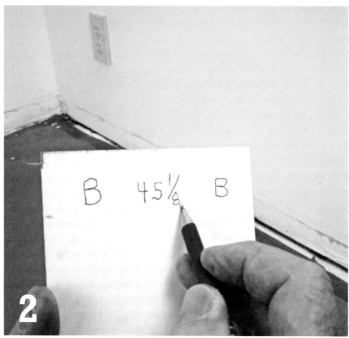

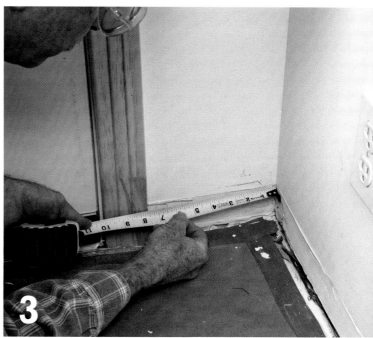

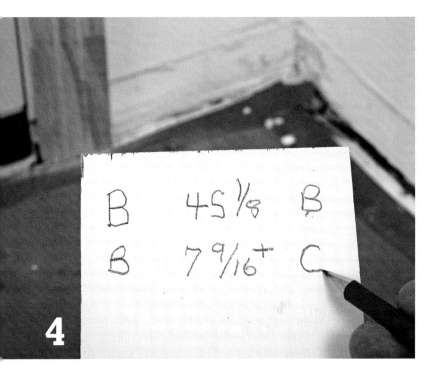

4. **Measure to 1/32 in.** The next piece measures 7 9/16 in. The "+" sign really means an extra 1/32 in. The left end is a butt cut; the right end is a cope (mark with the letter C), which means it must be mitered first.

5. **Measure long pieces a little long** Add 1/16 in. for walls over 8 ft., and add 1/8 in. for walls over 12 ft., then bow and snap the pieces in for a tight fit.

6. **Continue cut list** This piece is butt cut on both ends because the inside corner on the left will be covered by a cope cut on the next piece.

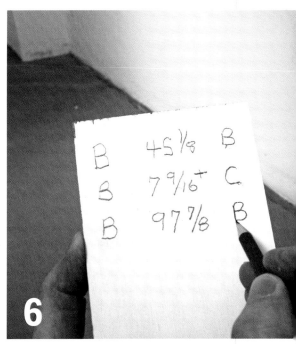

BASEBOARD | 59

Measure for the outside corners

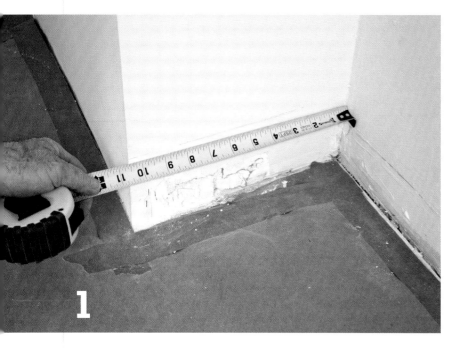

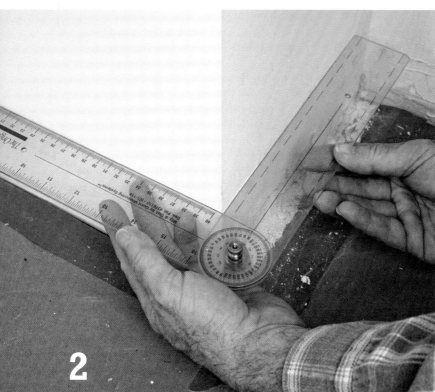

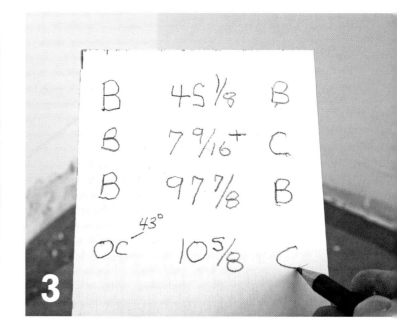

1. **Measure precisely** To avoid drywall mud buildup near the floor, always measure near the top of the molding—if necessary, trace a faint line across the top of the molding. For pieces up to 8 ft., measure outside corners precisely; for pieces longer than 12 ft. you can add a little if the molding is thin and flexible.

2. **Use a protractor** Outside corner miters must be cut at precisely the correct angle. Use a protractor to read the corner angle. Divide the corner angle in half to get the miter angle. The corner shown here corner measures 86 degrees—it's really out of square, which isn't unusual at all.

3. **Mark cut list** The baseboard piece here has an inside corner on the right, which gets coped, and an outside corner on the left (mark with the letters *OC*). The outside corner needs a 43-degree miter (86 ÷ 2 = 43 degrees).

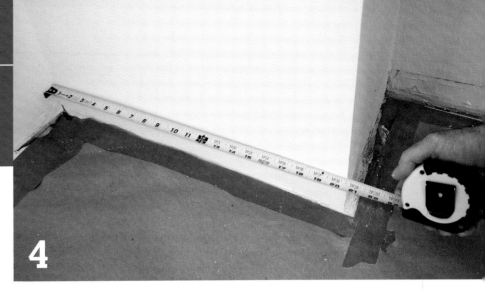

4. **Measure next wall** Be sure to hold your tape measure near the top of the molding and measure to approximately the same height on the outside corner. Avoid having to bend your tape by placing the butt end on the inside corner.

5. **Continue cut list** This piece has a butt cut on the left, because it's the first piece into the inside corner, and an outside corner (OC) on the right. Always write the measurement in the center of the cut list.

6. **Measure one room at a time** Too many pieces on a cut list will make the measurements and corner notes too small and difficult to read. You may end up mixing up the pieces. This wall is the most common type found in homes. It has two inside corners. The right corner will be coped, and the left will have a butt cut.

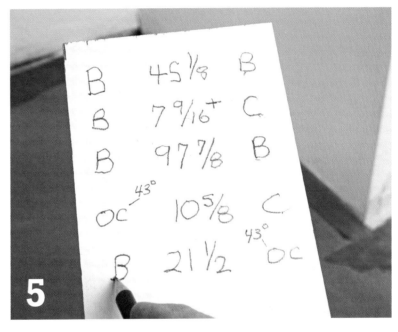

Cut tall baseboard

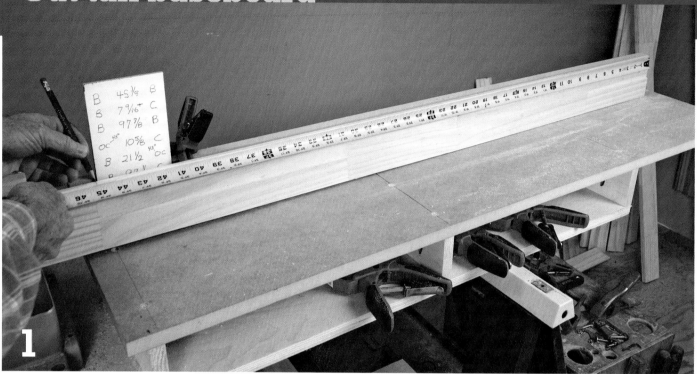

1

2

3

1. **Pull tape across first piece** The first piece has a butt cut on the right and measures 45⅛ in. Cut the right end square, and hook the tape measure there.

2. **Make fine lines** Use a sharp no. 2½ pencil and make fine lines, right on the measurement mark.

3. **Make two false cuts** Make the first cut wide of the pencil mark by at least an ⅛ in., just to locate the blade on the molding. Until you have more experience, make a second false cut, this one about 1/16 in. from the measurement mark.

4. **Cut to center of pencil line** Hold the material at the end of the miter saw fence and use your thumb to creep the pencil line up to the blade.

5. **Start second piece** The second piece gets a cope cut on the right side. Start by cutting an inside corner. For an inside corner, the long point of the miter must be against the fence. Once the small piece of waste is removed on the right side of the blade, the long point of the miter will be against the fence.

6. **Hook tape on long point** Inside corners are easy to measure because you can hook your tape measure on the long point, just as I've done on the right end of this piece. Before coping the inside corner, always make a measurement mark on the other end.

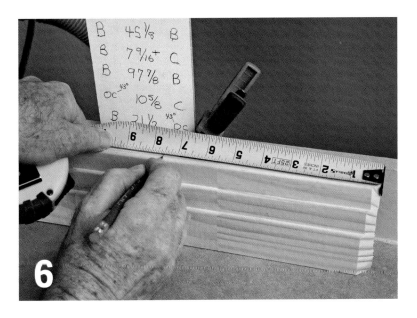

Head back to the saw

1. **Cut wide of mark** Some instructions are worth repeating. You'll hear this one several more times: Always hold the material at your miter saw with your free hand locked against the edge of the miter saw fence.

2. **Creep up on measurement** Now isn't the time to cut this piece too short—not after the cope has been cut. Use the thumb on your free hand to creep the measurement mark slowly to the blade. (See the DVD for coping.)

3. **Cut long piece** The next piece measures 97⅞ in. and has a butt cut on both ends. Having a miter saw with continuous extension wings makes it easy to cut both long and short pieces of molding.

4. **Cut inside corner and measure** The next piece has an inside corner on the left that gets a cope cut (C). Make that miter first, then hook your tape measure on the long point of the miter and make a measurement mark at 10⅝ in. for the outside corner (OC).

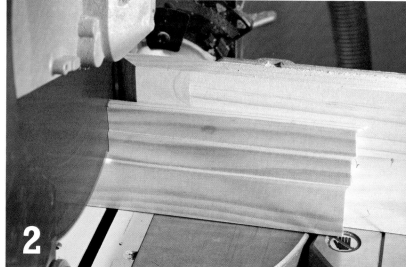

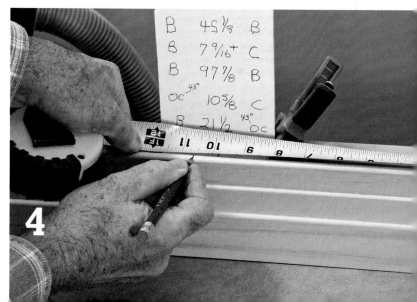

Cut the outside corners

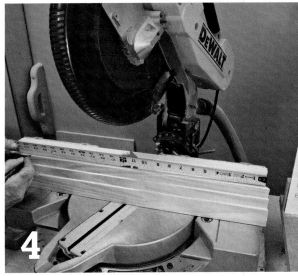
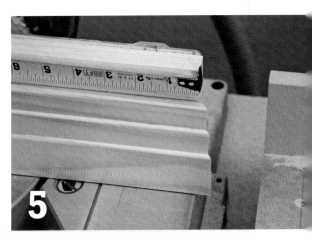

1. **Cut wide of outside corner** The piece that I've just coped on the right end needs an outside corner miter on the left end. Remember, for an outside corner, the short point of the miter must be at the back of the molding against the wall and your miter saw fence.

2. **Creep up on measurement** By coping the piece first, I'm able to cut the outside corner miter on a longer piece of material, and hold the molding comfortably with my free hand.

3. **Cut opposite miter** I always cut the right-hand ends first because I always move the material through the saw from my left toward my right. That way, it's easy to remember which way to swing the saw for the second piece—mating miters are always cut in the opposite direction.

4. **Never burn an inch** Measuring from an outside corner isn't as easy as an from inside corner—there's no long point on which to hook your tape measure. Some carpenters align the 1-in. mark on their tape measure with the short point of the miter. Don't do that. You'll forget to add that 1 in. back into the measurement.

5. **Measure outside corners** There's one simple and foolproof technique for measuring outside corners: Flush the short point with the edge of your miter saw fence and hook your tape measure on the fence.

Install on the short walls

1. **Fit tight but not too tight** Short lengths that butt against casing should not be forced into place if they're cut too long. Take the piece back to the saw, and trim it if it's more than 1/32 in. too long, otherwise you might crack the casing or even move the door jamb.

2. **Use a block** A snug fit is best. Don't hit the molding with a hammer, instead use a short block of wood to nudge the molding into position. Often a little drywall mud built up in the corners is all that prevents a well-measured piece from fitting on the wall.

3. **Trace overlapping copes** I cut my copes with an overlapping miter. Rather than make that overlap paper thin—because it often breaks off—I prefer to cut it 1/8 in. thick. With the piece in position, I follow the angle of the miter with my utility knife and score a line in the previous piece of baseboard.

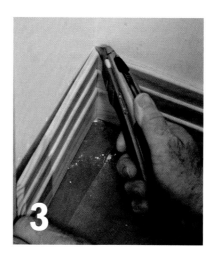

4. **Notch overlapping copes** Once the angle of the miter is traced, set the utility blade on the line and wiggle it up and down, cutting the angle a little deeper. When the blade is about 1/8 in. deep, twist it sideways, and the waste will snap out of the way.

5. **Position cope** Now isn't the time to lose patience. Don't force the cope together, otherwise the overlapping miter might snap off. Use a 5-in-1 tool to lift the molding and align the cope joint before pushing the piece against the wall.

6. **Tap into place** With the cope aligned, tap the back of the baseboard into position against the wall. A well-measured piece should be snug but not move the casing. The cope joint should close up almost watertight.

Install on the long walls

1. **Nail off baseboard at casing** Before pressure-fitting a long piece into position, securely fasten the butt end against the casing. That's the best way to prevent the casing from shifting.

2. **Snap in long pieces** Hold the center of long lengths away from the wall. Push the corner in, then remove your hand, and let the molding snap to the wall.

3. **Mark for overlaps** Butt-cut ends do not have to be removed for notching. Just follow the miter with your utility knife and wiggle the blade in about 1/8 in. deep.

4. **Make notches deep** Don't worry about cutting a notch too deep.

5. **Cover notches** A tight-fitting cope joint will always cover the notch.

Miter the outside corners

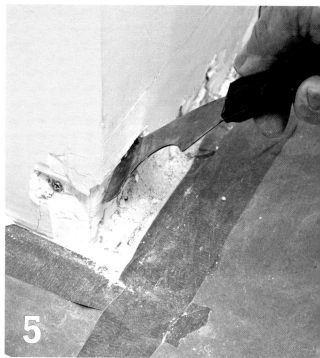

1. **Test-fit every piece** Always test-fit outside corners. Check that the cope fits perfectly, then make sure that the short point aligns exactly with the turn in the corner. Be prepared to trim a hair off the miter if necessary.

2. **Mark pieces in place** Some carpenters intentionally cut outside corners long, so they can mark them in place, without having to use a tape measure. Reverse the molding, so the long point of the miter is touching the inside corner, then trace a line along the outside of the previous piece of baseboard.

3. **Solve fit problems** I've heard carpenters say that 90 percent of finish work is solving problems. If the miter won't close, then find the problem. In this case, I can't press the molding tight enough to the wall because of the built-up drywall mud near the floor.

4. **Scrape problems away** Use a 5-and-1 tool to scrape away drywall mud, especially from inside corners, that might interfere with fitting a nearby outside corner.

5. **Carve problems away** If corners have too much buildup, mark the top of the molding, then drive a 5-and-1 tool into the wall $1\frac{1}{2}$ in. to 2 in. lower than the baseboard. Carve out as much of the corner as necessary to make a tight miter.

Fasten the baseboard

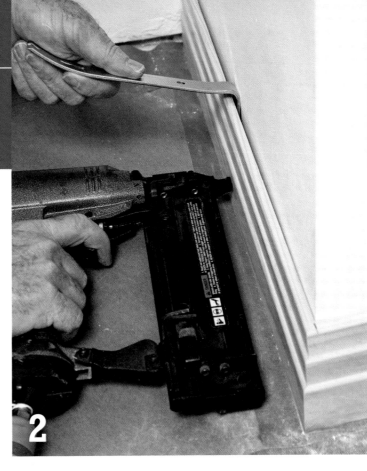

1. **Fasten outside corners first** Some corners must be fastened as you fit the pieces. Long pieces that butt in to casing are one example; outside corners are another example. I use glue and 23-ga. pins to secure outside corners.

2. **Fix bowed wall** Don't always nail off baseboard tight to the drywall. If the wall is bowed, hold the molding out with a small prybar, and create a straighter line. Don't attempt to make the line perfectly straight; instead, if the bow is bad, split the difference. Fasten the baseboard securely, use shims behind the bow if needed, and caulk the top at the wall. Most bowed walls will look fine using this method. For stain-grade molding, the wall must be skim-coated, or floated, with drywall mud to remove any bows.

3. **Fit joints first** Don't nail off any piece permanently until the joint is made with the next piece. You never know when you might have to lift a piece of baseboard just a little to get a perfect fit.

4. **Nail into studs** Frequently, baseboard can be nailed into the bottom plate by angle-driving nails near the floor. But the molding must be secured near the top too. Use a stud finder to find solid backing in the wall, and drive a nail at least every 32 in. on center.

CHAPTER
5

Chair Rail

Chair rail adds a defining traditional touch to any room by dividing the wall space while simultaneously punctuating the corners and window and door casings. Depending on the design, chair rail can be simple or it can be a challenge. Because this book is meant to be learning experience, I've included a style of chair rail that will build on the techniques I've already demonstrated, introduce a few new ones, and better prepare you for the next chapter on crown molding.

The design shown here could be improved by using three pieces of molding—a flat piece of 1×4 could be installed beneath the stool, and the cove molding installed on top of that. The overall scheme would be more reminiscent of stile-and-rail paneling. But the task would be a third more difficult. I've chosen instead to keep this chair rail simple, but not necessarily easy.

A two-piece chair rail has more punch than a single piece of molding, and it doesn't take much longer to install.

Planning chair rail

Installing chair rail is almost exactly like installing baseboard, except for two things: There is no floor so you need to establish a level, straight line around the room and you don't have to work on your knees.

Order of installation

The cove molding is installed under the chair rail, and most carpenters would probably install the chair rail first, but it's actually easier to install the rail after the molding has been fastened to the wall—just set the rail on the cove and toenail into the studs. And when you install the cove molding before the chair rail, you don't have to bend over so much.

TOOL BOX

Except for a couple of levels and the handsaw, there's no need to rush out and buy all the tools listed here, though they will make the job easier. At your local home center or lumberyard, pick up enough stool or chair rail and cove molding for your entire project—plus some extra. Sometimes it pays to spend a few dollars more just so you can practice. In this project I ripped beveled stool molding from wider stock. If you don't have a tablesaw, arrange for your home center to rip the material for you.

TOOLS AND MATERIALS

- Levels for drawing straight, level lines
- Laser level for shooting straight, level lines quickly and accurately
- Tripod for mounting laser level and adjusting height of laser line
- Starrett Tools protractors (one small, one big) for measuring corner angles with friendly miter saw angles
- Japanese saw for cutting notches in the chair rail
- Stool for the chair rail to match the stool and apron under the window
- Cove molding for trimming the bottom of the chair rail and creating a two-piece chair rail
- Caulking gun and caulk for applying molding to walls where nails can't be used and for filling in small cracks or gaps between the molding and wall
- Sandpaper and sanding block for sanding chamfers and joints
- Pin nailer to secure corners

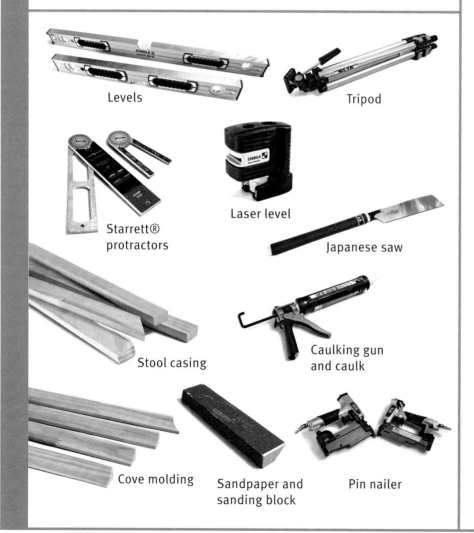

Levels

Tripod

Starrett® protractors

Laser level

Japanese saw

Stool casing

Caulking gun and caulk

Cove molding

Sandpaper and sanding block

Pin nailer

CUT UPSIDE DOWN AND BACKWARD

Cove molding, like crown molding, must be cut upside down at the miter saw. A good understanding of miters makes the task easy. Just remember what you've learned from previous chapters: The long point is always against the fence for an inside corner and the short point is always against the fence for an outside corner. Here's how it goes:

- If you try to cut cove molding right side up, the molding will be difficult to hold and secure at the miter saw, and you'll need to move your fingers closer to the blade. Besides, if the molding moves even slightly at the saw, the miter angle will be off.

- Cut upside down. Mark the top of your accessory fence with a large *L* on the right side and a large *R* on the left side. That will remind you which corner on your cut list you're cutting. Always move the material from your left toward your right. That way, you'll always cut the left-hand corner first.

Outside corners

Chair rail is a little more demanding than baseboard because the joinery is closer to the eye. Plus, with this profile, it's easier to use miters on inside corners. That means the corner angles must be measured accurately. Rather than cutting test pieces to identify the miter angles, learn to use a protractor. Your work will go faster and be more fun.

Cutting the parts

In Chapter 4, I used baseboard to demonstrate the fundamentals that apply to cutting all moldings. When we worked with a cut list, we looked at the whole piece, both corner cuts and measurements. Here, I demonstrate how to cut molding upside down. When the molding is upside down, the best way to avoid confusion is to approach each piece on the cut list as having three separate entries—left-hand corner, measurement, and right-hand corner.

Cutting the chair rail

While at the saw, save time and cut the chair rail too. That way, you can move back to the room with all the pieces at one time. Use the same cut list as well. Just remember to add extra so the chair rail can overlap the casing. The amount of overlap depends on the style of the casing.

The bevel

As you've come to learn, some things bear repeating. The most common mistake when cutting the chair rail is cutting the bevel or chamfer on the butt ends in the wrong direction. Remember that the top of the chair rail is the long point of the bevel. This molding mimics the dado molding on classical columns. Water dripping off the molding would fall straight off the top edge, without dripping down the bevel.

Installation

Cove molding is small and thin, making it flexible and forgiving. A little too tight is always better than a little too lose. Spring the center away from the wall, fit the ends, then snap the molding in place.

Chair rail isn't nearly as forgiving or flexible as cove molding. And the miters must be watertight—each and every one. Don't lose patience, don't rush the job, stay focused, and enjoy the challenge of fitting each corner. Once you learn how to work with your tools in concert, especially the protractor, measuring tape, and saw, many of the pieces will go right in. Some will still need a little fiddling. Make that fiddling fun.

SIMPLE CHAIR RAIL DESIGN AND JOINERY

I chose a two-piece chair rail for the project in this book because it introduces a couple of interesting challenges. You could use a simpler one-piece molding, which installs much more easily. But a two-piece molding has more depth, casts a deeper shadow line, and provides more drama in a room. Plus, it's more fun to install.

The chair rail molding shown here is notched and overlaps the casing, which is the preferred method for terminating chair rail that is proud of the casing. This method is also the best solution if the wall has wainscoting and the cove molding is installed on top of the top rail. If the cove molding is flush with the casing, add a self-return at the back of the casing. Then overlap the chair rail onto the casing.

This two-piece chair rail molding is notched and overlaps the casing.

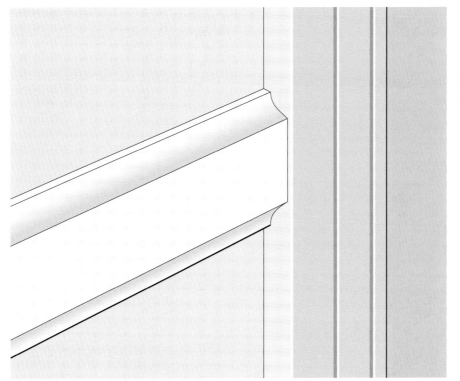

DO THIS

One-Piece Chair Rail

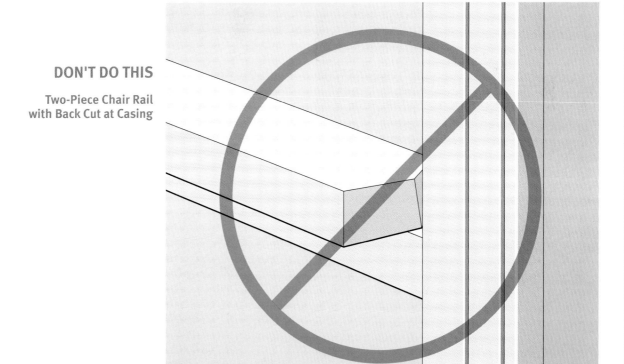

DON'T DO THIS

Two-Piece Chair Rail
with Back Cut at Casing

CHAIR RAIL | 75

Establish a level line

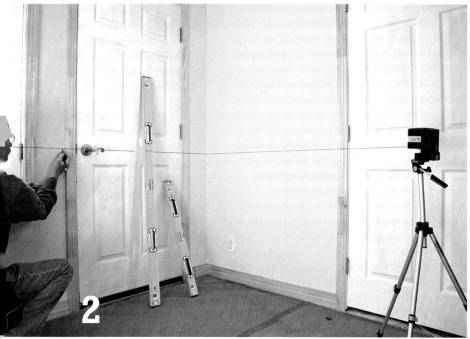

1. **Use levels** Pick a height that suits your room. Chair rail can be installed at almost any height, from 28 in. to 72 in. In many homes, the windowsill elevation is also the chair rail height, so the stool design wraps around the entire room. In this room, I'm placing the chair rail so it aligns with the lock rail on the doors.

2. **Use laser on tripod** Lasers are the quickest way to run a perfectly level line around a room. Without a laser, you'll need several levels in various lengths or one small level and a straight board that you can cut down for shorter walls.

3. **Adjust height** I mount my laser on an inexpensive lightweight camera tripod, so it's easy to move around and easy to adjust the height of the line. Once the line is set, I mark a pencil line next to the casing beside every door and window and in every corner of the room.

Make a cut list

1. **Butt cove molding to casing** The design I'm using works well if the cove molding is the same thickness as the casing, as shown here.

2. **Miter inside corners** On tight corners, cut down a True Angle® protractor, or use a 5-in. Starrett Tools® Prosite™ protractor, which reads angles that match most miter saw gauges.

3. **Measure carefully** I use a single cut list for the cove molding and the chair rail. For butt joints, I mark my cut list with a *B*; for inside corner miters, I use a *V*.

4. **Measure every angle** A full-size Prosite protractor is easier to read than a plastic protractor but much more expensive. The Prosites have angles that match a 0-degree miter saw gauge.

Measure for outside corners

1. **Check outside corners** Read the angles to within 1 degree, then cut the molding to a ½ degree. To get an accurate measurement, use the longest protractor you can. Measure the angle right at the chair rail height. Often the angle of a wall will vary from the baseboard to the chair rail.

2. **Measure long** If you're not certain of the measurement, err on the side of measuring a little long rather than a little short, then trim long pieces to fit perfectly.

3. **Mark outside corners** The cut list for chair rail should look just like one for baseboard, except the inside corners are marked with a *V* for a miter rather than a cope. Outside corners are still marked *OC*. If you're installing one-piece reversible rail, then cope the inside corners.

4. **Use a laser measure** Measuring chair rail on long walls is harder than measuring baseboard because the tape measure must be held up in the air. A laser measure speeds up the task considerably and guarantees accuracy.

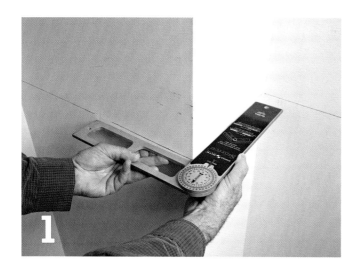

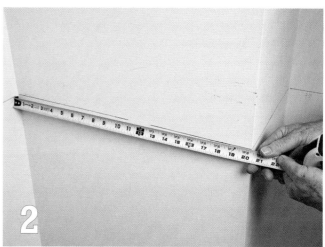

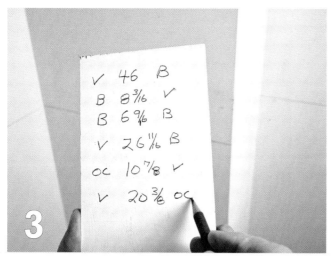

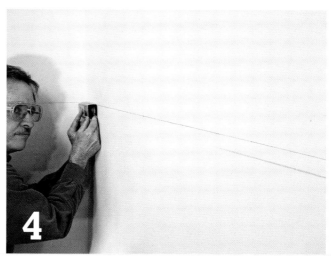

Cut the cove molding

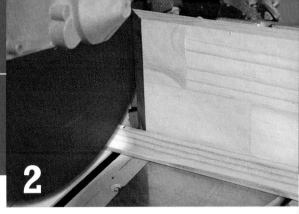
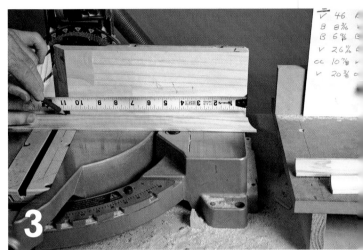

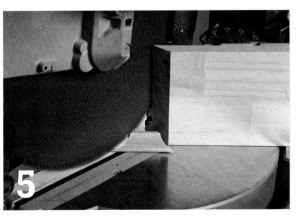

1. **Cut left-hand corner first** Because the molding is upside down, the left-hand corner is always cut first. Hook your tape measure on the long point and make the measurement mark.

2. **Cut right-hand corner second** Now is the perfect time to practice cutting crown molding, because making mistakes on cove molding is much less expensive. The right-hand end is butt cut. Creep up to the measurement mark.

3. **Flush short point** Outside corners have the short point against the fence. Flush the short point of the miter with the edge of the fence, and hook your tape measure on the fence. The piece in the photo is number 5 on the cut list. Notice that the outside corner is on the left-hand end of the molding, which is upside down.

4. **Don't swing saw** For pieces with an inside corner on one end and an outside corner on the opposite end, you don't have to swing the saw. Just slide the piece over and creep up on the measurement mark.

5. **Swing saw** If you always move the material in the same direction, from your left toward your right, remember to swing the saw in the opposite direction for each miter.

Cut the chair rail

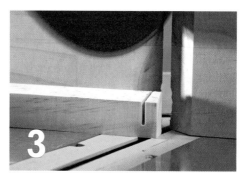

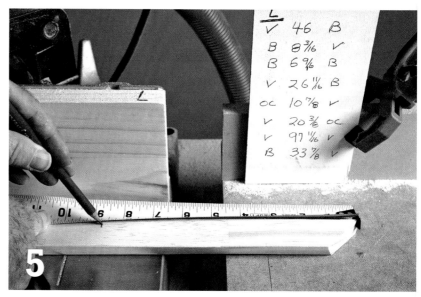
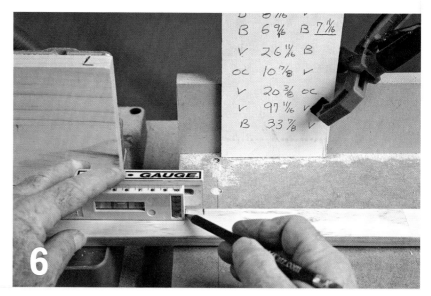

1. **Make notches easy** Make both cuts (the shoulder and the depth) for the notches the same dimension. Set up a gauge block or Trim Gauge exactly the thickness of the casing.

2. **Cut first end** This first piece has a butt cut on the right end and measures 46 in. to an inside corner. This material is cut right side up, so the right end is the right end.

3. **Watch beveled edge** Just like the window stool, the chair rail must be chamfered 15 degrees on both ends. The long point of the chamfer is always at the top of the molding, which in this example is against the fence.

4. **Creep up on marks** With the beveled edge pointing toward you, the measurement marks for the inside corner are against the fence and thus are difficult to see. Creep up on those marks carefully.

5. **Add 9/16 in. for overlap** Don't forget the overlap where the chair rail meets the casing. According to the cut list, this piece (the second in the list) measures 8 3/16 in. I added 9/16 in. and thus made the measurement mark at 8 3/4 in. You can add the figures in your head or cut a small piece of casing and use it as a gauge.

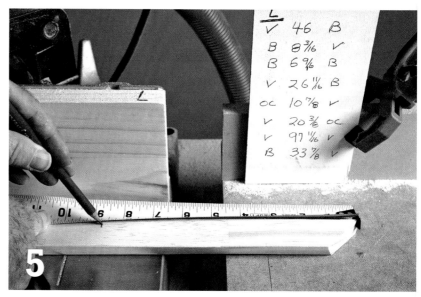

6. **Use a Trim Gauge** You can use the Trim Gauge to add the 9/16 in. Align the edge of the gauge with the measurement mark, then mark the molding at the shoulder of the gauge. Using a gauge is much easier than doing the math.

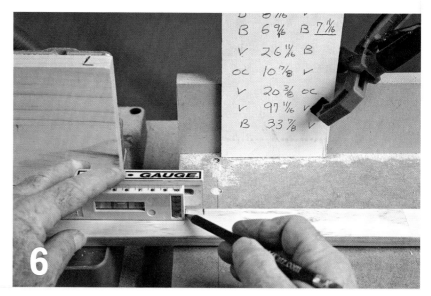

80 | TRIM MADE SIMPLE

Keep your eye on the bevel

1. **Roll material** Measurement marks are made on the top of the chair rail. To see the marks, roll the material so the top faces you, whether the bevel is up or down. Here the bevel is against the base of the saw and the fence.

2. **Mark notch depth** Use the Trim Gauge to mark the thickness of the casing, or the depth of the notch.

3. **Mark notch shoulder** Use the Trim Gauge to mark the overlap, or shoulder of the notch. Make sure your pencil is sharp.

4. **Cut notch with jigsaw** If you have a jigsaw that runs without too much vibration, use that for cutting the notches. Clamp the molding securely to your workbench, then hold the saw with your left thumb on the edge of the base. Use your left index finger to guide the cut, riding like a fence along the molding. Keep your other fingers wrapped in your fist.

5. **Use a handsaw** A Japanese pull saw works great for cutting small notches. Japanese saws are sharp and cut on the pull stroke, so it's easier to follow a line and make delicate notches.

6. **Clean up with a rasp** If the cut isn't square or the piece is a little too tight, trim the notch using a Shinto saw rasp.

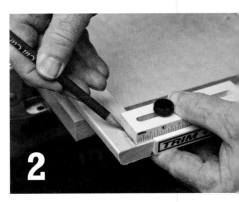

Install the cove molding

1. **Make cove molding flush** Before installing any of the cove molding, be sure that it's flush with the casing. A couple of passes with a sharp low-angle block plane should be all that's needed to make a perfectly flush joint.

2. **Check fit** Check each piece before gluing the corners. Most often, the pieces are a hair long, which allows for a little trimming if the miters aren't aligned perfectly.

3. **Use a stud finder** Carry a stud finder to locate each stud in the wall, and be sure to fasten the cove molding securely to the wall. Use 18-ga. 2-in. brads. That way, the cove won't slip down the wall while you're installing the chair rail.

4. **Use a sanding block** Carry a block of wood wrapped with sandpaper—you can glue the sandpaper on to the block with 2P-10 adhesive or secure it with staples. A sanding block is perfect for touching up miter angles right at the corner.

5. **Pin-nail outside corners** Glue and fasten outside corners with a 23-ga. pin nailer. If you don't have a pin nailer, use ¾-in. brads, but clip the points before driving the nails so the molding won't split.

6. **Use adhesive caulking** Caulking is also great for fastening the molding to the wall, especially if there are no studs or backing available. Apply a tight bead of adhesive caulking (DAP® latex caulking acts as an adhesive). Secure the molding temporarily with brads until the adhesive dries.

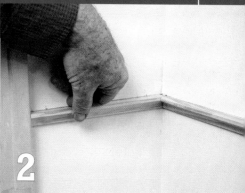

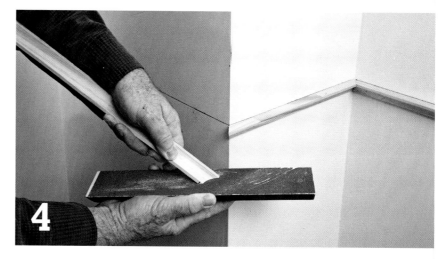
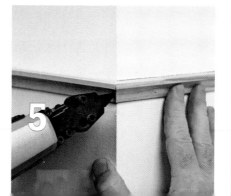
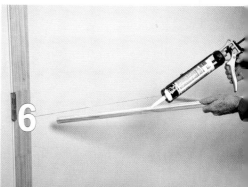

Install the chair rail

1. **Fit all pieces first** Check the fit on all contiguous pieces before fastening any of them. That's a lesson most carpenters learn several times, sometimes in the same day. The long piece on the wall is not fastened. It fits snugly between the casing and the corner.

2. **Toenail chair rail** After fitting every joint (on the run shown here, there's only one corner but two joints to check), glue all miters and butt joints, then fasten the chair rail to the wall using 18-ga. 2-in. brads or 15-ga nails, if necessary. Use a stud finder to be sure the rail is well secured. On walls without proper backing, apply a bead of adhesive caulking to the cove molding first.

3. **Check outside corners** Don't be surprised if several pieces require a little trimming. Fit as many pieces as possible at one time, so you'll make fewer trips to the saw. Once the fit is fine, glue both sides of every miter.

4. **Fasten miters in both directions** Use 23-ga. pins or 3/4-in. finish nails to fasten outside corners in both directions, driving the nails right through the bevel.

5. **Detail with sanding block** Sand corners smooth and flush with a sanding block. Be careful not to sand across the grain on the top of the chair rail. On the front edge, hold the block at the bevel angle. If one miter is a little longer, a sanding block will clean up the joint quickly. For small gaps, spread a little extra glue in the joint, then lightly sand over the glue, filling in the joint with glue and sawdust.

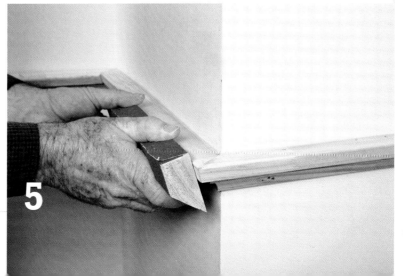

CHAPTER 6

Crown molding

Finally we come to the star of the show, crown molding. For a number of reasons, crown molding is the topic of the last chapter. Remember, I started at the floor and worked my way up the wall because that's the way a house is built. Second, I began with the easiest project—cutting casing—because those techniques provide the fundamental skills you need to tackle the types of trim that came next.

The level of skill needed increases as you climb higher up the wall. Now that we've reached the ceiling, it's time to celebrate, and that's another reason I saved crown for last. Installing crown molding is the one of the most enjoyable parts of any job, when effort and expectation meet in perfect harmony.

In this chapter, I demonstrate how to use all the lessons from previous chapters, plus a few new ones, to conquer crown molding.

Crown molding does the most to dress up a home. In fact, most traditional homes seem unfinished without crown molding.

Cutting crown molding

In the last chapter on chair rail, I demonstrated why cove molding must be cut upside down and backward. Crown is no different from cove molding, except it's bigger and it has no back, which makes it a bit harder to work with. But don't let that scare you. In this chapter, I show you foolproof techniques that make the challenge easy for anyone who has read and practiced the techniques presented in the earlier chapters. If, however, you jumped to the end . . . shame on you!

Crown molding jigs

Trust me on this: The jigs I suggest you make are real problem solvers. Take pride in your shopmade jigs, fixtures, and accessories; these are tools you can't buy but that make a world of difference in your work. The crown jig is simple to make and will allow you to lay out and install crown molding like a professional the first time.

TOOL BOX

If you've followed all my tool recommendations, starting with Chapter 1, you've reached the break-even point: You don't have to buy any new tools for this chapter, except a ladder. Now that you're an experienced carpenter, you'll have to make your own tools—these cannot be purchased at a store. Crown molding requires four jigs every carpenter should carry in his or her toolkit. The crown jig must be made to fit each specific molding.

TOOLS AND MATERIALS

- Crown jig for laying out and installing crown
- Coping jig for coping crown with a coping saw or a jigsaw
- Lay-up table for preassembling outside corners and mockups
- Crown clamp for installing crown alone
- Crown for trimming the ceiling

Crown jig

Coping jig

Lay-up table

Crown clamp

Crown molding

Measurement and layout

Crown molding sits on the wall and angles up to the ceiling, so it's not easy to measure at the ceiling. Because crown is cut upside down at a miter saw, measuring the bottom of the crown where it sits on the wall is the easiest, fastest, and most accurate method. If you measure crown at the bottom, where it sits on the wall, when you turn the molding upside down at the miter saw, you can make all your measurement marks on the bottom, and guide the sawblade right to those marks.

When I installed baseboard, I entered the room and moved toward the left so that cope cuts were always made on the right end of each new piece. Because crown molding is cut upside down and because it's easier to cope the left end, always move toward the right when installing crown. Also, make sure you cope

MAKE A STOP FOR CUTTING CROWN

Understanding some basic guidelines for cutting crown molding will help prevent mistakes and lower your risk of injury. Many a beginner has learned these lessons the hard way.

First, don't try to cut crown by holding it upright, mimicking the way it's fastened to the wall. The material will move as soon as the blade gets near it, and you won't make accurate cuts. Cut crown molding correctly by flipping it upside down so it "nests," or props, against the saw base and the fence.

Using a crown stop is the only way to cut each piece to a precise length and angle with few mistakes. Start by measuring the projection of the crown across the base of the saw (which is how the molding rests against the ceiling). Next, cut two spacer blocks off a 1×4 the same length as the projection. Save the spacers so you can set up the saw exactly the same way later. Place a spacer against the saw fence, then clamp a sacrificial piece of 1×6 in front of it.

Measure the projection of the crown across the base of the saw, which represents the ceiling.

Place a 1×4 spacer against the saw fence, then clamp on a piece of 1×6.

You can also rip the spacer from a simple piece of stock.

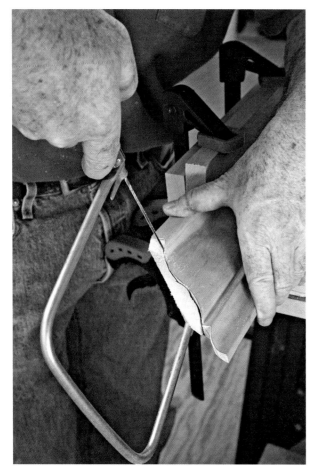

The companion DVD shows you how to use a coping saw or jigsaw to cope crown molding.

USING A COPING SAW

Working with a coping saw is a bit tricky, but a coped miter can also be cut with a jigsaw. But if using a jigsaw upside down makes you uncomfortable, use a handsaw. Most crown profiles are small enough so that they're easy to cut, especially if you make a coping jig. Just be sure to use a sharp, new blade. You can mount your blade so that it cuts on the push stroke or the pull stroke. I prefer to cut on the pull stroke, though some carpenters prefer the push stroke because the face of the molding isn't frayed by tear out. (To see exactly how I use the coping saw to cut a coped miter on crown molding, see p. 97.)

the first piece, even though the previous piece of molding isn't installed yet. Coping the left end is easier and nearly every piece will be cut the same way. (To see how I cope the crown molding, see the companion DVD.)

Cutting accurately

Making a cut list simplifies cutting crown considerably. Just remember, you cut crown molding upside down, so the corners are reversed. Don't try to remember both corners and the measurement mark. Look at the cut list, see the first left-hand corner, and cut that joint. Then look for the measurement length, mark that, and finally check the cut list for the right-hand end. If you move the material from your left toward your right, you'll always be cutting the left-hand corners first. I put an L above the left-hand corner column of my cut list. The other advantage to flipping crown upside down in a miter saw is having the measurement marks facing up, so you can guide your sawblade right to the mark.

Now we do it all again. For the most part, every piece of crown molding is cut the same way, unless you have an inside or outside corner. The last piece here has an inside corner on the right end, so this piece must have a cope on the left end.

Installation

Installing crown molding, if it's cut correctly, isn't that difficult; but there are some basic tricks that make the job even easier. Like most finish work, try to see the whole picture before you begin. Look at every corner and imagine what you'll be doing at each one. Always have a plan for how you're going to finish the room, know which will be the first piece and which will be the last piece you install.

Long walls

Just because you started measuring at a specific wall, doesn't mean you have to start installing molding on that wall. But it is easier to start installing the pieces on a long wall—a piece of molding long enough to fasten securely to the wall in the middle and still bow or bend at the corners.

Outside corners

Every carpenter loves installing crown on outside corners, especially if the corners are square and the ceiling is flat. But even in poor conditions, with the right preparation, your joints should line up pretty closely. If they don't, there's usually enough room on one of the pieces to trim the corner a hair and make a perfect fit. Just be patient. Don't hurry. This is the best part of the job.

Securing the molding

Until most of the pieces are up on the ceiling, I keep all nails at least 2 ft. away from the corners. That way, I can tap one piece of crown up a little or the other down a bit to get a perfect fit. Once I have all the pieces in place, I fasten every corner I can, while I'm there.

Finishing the installation

My strategy with this room is probably clear to you now. I started with the nearest long wall to the outside corner, so I could get that corner out of the way first and fasten everything permanently once those pieces were in place. Crown molding is a lot easier to install than people think; and you can probably start to agree by now. Most of the pieces are identical, except for outside corners and strange walls. That's why I try to tackle those parts—the hardest parts—first.

MAKE A CROWN HOOK

Installing crown molding is often a two-person job, but there are times when carpenters install crown alone. It's not that difficult to do if you have a crown hook. You can make one of these tools yourself, with just a $1/8$-in. by 2-in. strip of aluminum, a few drill bits, and a hammer.

- **Drill a keyhole.** Before cutting and bending the aluminum, drill three holes about $1/2$ in. from the end. Start near the end by drilling one smaller hole a little larger than the threads on a wood screw and a second hole below the first one that's a little larger than the first hole, and finally a $1/2$ in. hole at the bottom, so it's easy to slip the hook off a screw.

- **Cut and bend the hook.** Cut the hook 6 in. or 8 in. long. If you don't have a vise, clamp the aluminum to your worktable, and smack it with a hammer; then turn it over, and smack it some more. The clip doesn't have to be pretty. I just has to hold up the crown molding.

Make a crown jig

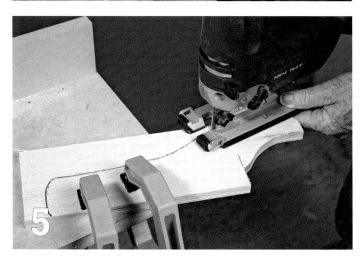

1. **Measure crown projection** Start with a piece of plywood about 6 in. wide and 12 in. long. Measure and mark the crown projection across the top of the plywood. To measure the projection, you can also use the blocks you cut for setting up your crown stop.

2. **Make a clean, straight line** Draw a sharp line at that measurement mark. Crisp lines and precise measurements are especially important when making jigs for repetitive work.

3. **Cut a 1-in. piece of crown** Place a small piece of crown on top of the jig, right on the line. Make the crown flush with the top of the jig and perfectly flush with the back of the jig too. Dawn a sharp line across the bottom of the crown.

4. **Connect lines** Draw a line connecting both the top and the bottom lines. You don't have to follow closely against the profile. The two sharp lines—one at the top shoulder and the other at the bottom foot of the crown—are all that really count. The rest is waste.

5. **Cut carefully along lines** So the crown fits snugly in the jig, cut the top and bottom lines with precision, cutting to the center of the pencil line. The remainder can be cut quickly. Also cut in a comfortable handle. And don't make the jig too long or it won't fit between the crown and the window and door casings. Guess how I know that?

Lay out for crown

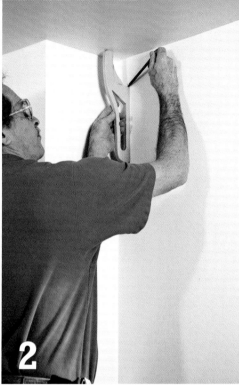

1. **Mark all inside corners** Use the crown jig to mark the bottom end of the crown molding at all inside corners. The crown jig will span the drywall, mud buildup, and any irregularities in the joint at the wall and ceiling.

2. **Mark all short lengths** Use the crown jig to mark the bottom of the crown on all short walls too, on both ends. Snapping chalklines is not necessary.

3. **Mark center of long runs** Make a few marks on the long runs too, just to make it easier to install the molding later.

4. **Mark outside corners with a mockup** Use a mocked-up molding corner to mark all outside corners; that way you'll know exactly what's going on with the walls and ceilings. If the wall is out of square, you'll know to check it with a protractor.

Measure the crown

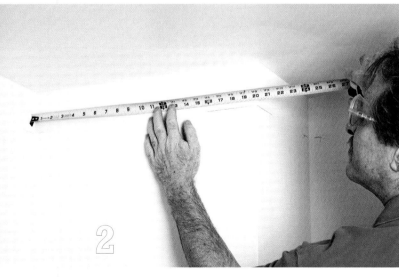

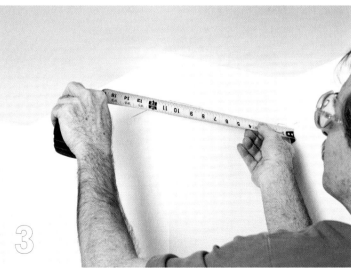

1. **Start on a long wall** If the first piece of crown is long, it's easier to bow the center and slide in the last piece. Installation can also start at an outside corner, though I prefer to fasten outside corners securely, as soon as I reach them—they're the focal point for crown molding.

2. **Measure carefully** The nice thing about coping all inside corners is that measurements don't have to be perfect, in fact, cut pieces a little long, especially on long runs, then spring the corners into place for super-tight joinery. The run shown here has an inside corner on the left and an outside corner on the right, so the measurement must be precise.

3. **Check corners with a protractor** I already know the corner is square because I checked it with the mockup. The molding has an outside corner on the left and an inside corner on the right.

4. **Make a cut list** Notice how most of the pieces have a cope on the left and a butt cut on the right. If a room doesn't have outside corners, all the pieces are the same, if you cope the left-hand end of the first piece too.

5. **Use a laser** Measuring long walls for crown molding is tough when you're working alone. One technique is to cut a piece of thin molding, like door stop, exactly 100 in. long. Place that piece into the corner, mark the end, then measure from the opposite corner back to the mark, adding 100 in. for the total length. This technique works for measuring baseboard, too. But these days, I use a laser. It's much faster and more accurate.

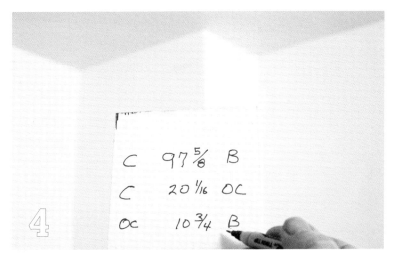

Cut the first piece

1. **Cut an inside corner miter** Cope cuts are always on the left hand end, even on the first piece. So always begin with you miter saw swung to your left.

2. **Cut tip off miter** The bottom of the crown can't be coped so it must be cut in a butt cut or an overlapping miter. For paint-grade molding like this crown, cut a butt cut at your miter saw. Cut the tip of the miter off, right to the short point on the top fillet. We're going to cut several, so you'll have plenty of opportunity to practice.

3. **Hook tape to long point** Mark your measurement at the back of the molding, where the short point or long point must be cut. Use a sharp pencil.

4. **Butt-cut opposite end** The opposite end of the first piece is, like most pieces, a butt cut. That's another reason why using coped inside corners is faster. Even though it's a butt cut, creep up on the measurement mark and cut precisely.

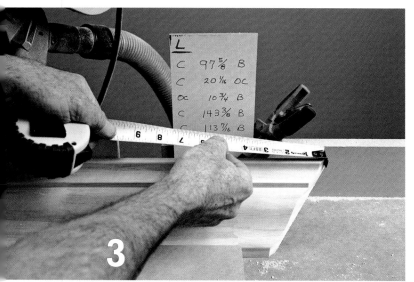

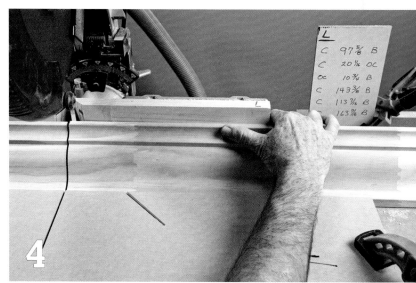

94 | TRIM MADE SIMPLE

Cut the next two pieces

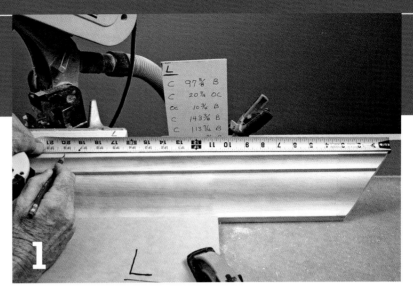

1. **Cut inside corner miter** Hook your tape measure on the long point, and mark the measurement.

2. **Make a creep cut** Don't try to cut right to the mark, because you might cut the piece a hair short. Cut wide of the mark, and then sneak up using small creep cuts.

3. **Keep free hand at fence edge** Never move your free hand in toward the blade because you won't have adequate control of the material, and you'll risk cutting yourself. Lock your fingers against the fence and use your thumb to creep the measurement mark toward the blade.

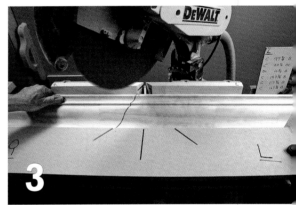

4. **Cut tip off miter** First make a cut wide of the short point on the top fillet. That will make it easier to position the blade exactly on the miter. The eased edge makes this tip on the crown molding a little harder to trim.

5. **Draw centerline** If you need to, draw a pencil line to the center of where the short point of the miter intersects with the bottom flat edge, or fillet, on the crown. Because the crown is upside down, that fillet is at the top in the photo.

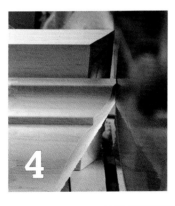

6. **Swing saw** Cut outside corners so the short point is against the fence. Then flush the short point with the edge of the auxiliary fence, hook a tape measure on the fence, and measure for the next cut.

Build a coping clamping station

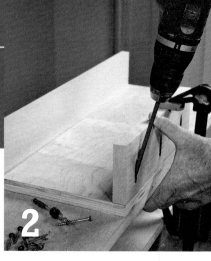

1. **Make the fence** The easiest way to cope crown, so that you don't waste time back cutting too much of the material, is in a clamping station, one that positions the crown just the way it sits on the wall. The clamping station fence must be adjustable, for different sizes of crown. Use a 12 in. length of 1x2 for the cleat, and a 12 in. piece of 1x4 for the fence. Clamp and fasten the two pieces together with screws.

2. **Make the base** A piece of ¾ in. plywood 12 in. long and 10 in. wide makes a fine base. Cut pocket holes in a 12 in. piece of 1x4, then clamp and fasten the 1x4 to the base with pocket screws.

3. **Locate the fence** Imagine the base of the saw is the wall and the fence closest to you is the ceiling. Position the fence so that the top shoulder and bottom foot rest flat on the fence and the base. Pencil a line at the back edge of the cleat.

4. **Fasten the cleat** Clamp the cleat on the pencil line. Pre-drill three holes for screws, then fasten the cleat and fence to the base. With the crown in this position, you'll have a perfect perspective of the joint: You'll be looking straight down at the miter as you cut the cope, from the same direction as the next piece of crown that makes up the corner.

5. **Secure the crown with clamps** Use a small piece of wood as a shim, about the length of the jig and wide enough so, when resting on the top of the crown, the shim sits above the jig. Clamp the shim, the crown, and the jig, at the same time, to a sturdy workbench.

6. **Darken the profile with a pencil** Use a sharp pencil and lay the lead flat on the edge of the miter to darken the profile, making it easier to see.

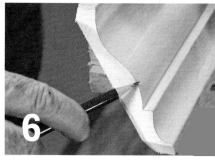

Coping with a coping saw

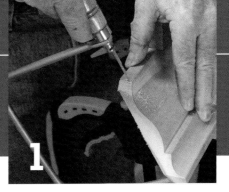
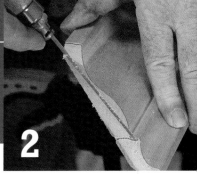
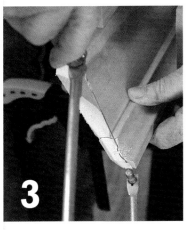

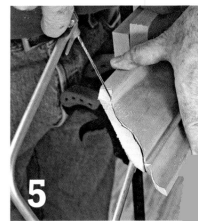

1. **Start by making relief cuts** Cut along the top fillet, tilting the saw just a little, so the edge will be back cut and come up tightly against the last piece of crown. Keep the saw moving in a steady rhythm. Don't push on the saw. Let it cut by itself and "will" it to follow the profile.

2. **Cut along other fillets** Every fillet needs a straight cut because it butt's into the previous piece of molding. The bottom of the crown is exactly the same, but we cut that fillet at the miter saw by nipping off the tip of the miter.

3. **Follow the lower cove** Starting the blade into the workpiece is easiest if you put very little pressure on the handle. Instead, concentrate on developing a smooth even stroke, then gently press the side of the blade into the material and "will" the blade to follow the profile. This is one technique where trying hard is *not* the right approach.

4. **Start with a gentle push** Here's another way to start a cut. Place the blade on the very tip of the fillet and give the teeth a gentle push, so they start to bite into the wood grain, than begin your stroke.

5. **Follow the upper S-curve** The cyma recta curve at the top of the molding is gentle and easy for the blade to follow. Pick up the pace along this cut by quickening the stroke, not by adding pressure on the blade.

6. **Clean up the cut** Use a piece of 80-grit sandpaper wrapped around and stapled to a sanding block. Odd scraps of wood, anything with a sharp corner, make excellent sanding blocks. For some crown profiles, with tight turns, rasps and files are a necessity.

Coping with a jigsaw

1. **Tip the jigsaw back** Start by touching the nose of the foot against the back of the molding, then slowly raising the motor until the blade contacts the molding.

2. **Tip the jigsaw forward** Once the blade begins to cut, follow the top fillet by raising the back of the motor and tipping the saw blade forward. That's the best way to control the depth of cut.

3. **Make relief cuts at every fillet** Everywhere the molding stops at a right angle needs a relief cut, otherwise, the blade will be trapped during the cut. Always start by tipping the blade back.

4. **Rub the blade against the wood** Don't be aggressive with the tool. Allow the blade to run while finding the right entry point, then lift the back of the motor and follow the fillet.

5. **Tip the saw sideways** The coping foot allows the saw to tip in almost every direction. To start a side cut, tip the saw sideways, run the blade without forcing it into the molding and allow the teeth to find the right entry point, then tip the saw forward.

6. **Don't push the blade** Let the saw do the work. Tip the saw back and forth to make the cut and follow the profile. Until you've practiced a lot, while tipping the saw blade forward, cut slightly away from the profile; while tipping the saw blade back, use the blade to "sand" right to the profile.

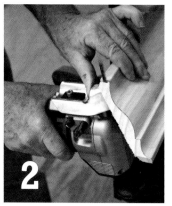
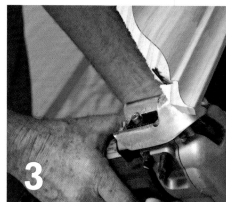
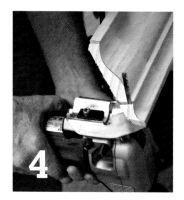
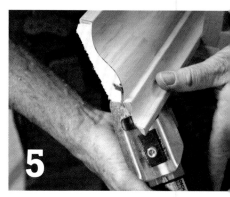

Start on a long wall

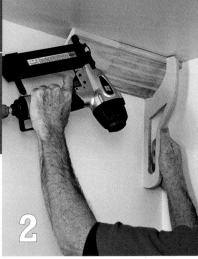

1. **Use a starter piece** Because the left-hand end of the first piece of molding is coped, you should start by tacking a short piece of molding to the last wall you'll be working on. That way, the first piece you install will be positioned perfectly for the last piece—not too close to the corner and not too far away either.

2. **Tack starter piece** The starter piece must be easy to remove later, so use only one nail at the top, near the very back of the molding.

3. **Center your ladder** If the wall isn't too long, position you ladder near the center of the wall, climb up, and engage the left-hand coped end against the starter piece. If the starter swivels down a little from the ceiling, just push it back up with the coped end.

4. **Use crown jig** Position the crown molding against the wall and ceiling with a crown jig, then nail the crown on both sides of the jig. If the molding is a little long, let the center bow out, poke the ends into each corner, then snap the center into place. Don't bother checking the corners or the position of the crown. The starter piece and the crown jig ensure a perfect fit.

5. **Don't nail ends yet** For the most part, don't fasten the crown at the ends or corners until all the pieces are up—that leaves some wiggle room for adjusting corners later.

6. **Take care with outside corners** The inside corner on the joint shown in the photo fits pretty well. But I'm using a 23-ga. pin nailer to tack the molding to the wall. The outside corner must be checked for fit before proceeding.

Install outside corners

1. **Fit pieces** Before applying glue to the joint, check the fit. Make sure the outside corner fits tight from top to bottom, then check the fit of the cope on the left end of the left-hand piece. Finally, check to be sure the crown is touching the wall. Sometimes these pieces end up a little long, which is a good thing because it allows for a little trimming for a perfect fit.

2. **Glue outside corners** Apply glue to both sides of the miter, then fasten the joint securely with 18-ga. 1-in. brads or 23-ga. 1-in. pin nails. I prefer pin nails because they can't been seen, and I can shoot a lot of them in both directions. This joint will never open.

3. **Use crown hook for long pieces** If you're installing really long pieces of molding by yourself, fasten a hook ¼ in. to ½ in. below the bottom line of the molding. I use a magnetized stud finder to look for a stud near the corner. Then I use my crown jig to locate the bottom of the crown and drive in a square-drive screw, leaving the head of the screw barely touching the hook, so it's easy to slide the hook up and out once the crown is in place.

4. **Position your ladder** Always plan ahead for where you'll want to climb up, especially when you'll be lifting the crown as you climb the ladder. Slide the crown into the hook before climbing the ladder.

5. **Raise crown over your head** If the hook isn't really close to the corner, raise the crown above your head and then slide the molding into the hook.

6. **Lift and fit crown** When using a crown hook, I position my ladder at the opposite corner from the hook, so I lift the crown and fit the first corner at the same time, rather than nailing the center and then moving my ladder to the corner. Like most pieces, the one shown here is a bit long. The center bows out, allowing me to pressure-fit the cope joint.

Fasten the corners

1. **Pressure-fit coped corners** When measuring crown, I make sure long pieces are a little long and the short ones are cut precisely the length of the wall. Notice on the corner shown how the molding to my right is bowed from the wall in the center. That's a good thing. That bow means the piece is a little long, and I can pressure-fit to the cope joint.

2. **Check position** Before fastening the middle, use a crown jig and check that the molding is in the right position. This will make the job look a lot better, and it will make assembling the next corner much easier.

3. **Slip out hook** The keyhole slot in the crown hook makes it easy to remove the hook. Don't worry about the screw. Leave that for the next carpenter to puzzle over.

4. **Adjust crown** Usually long pieces have to be adjusted to the layout lines marked with the crown jig. If the crown needs only a nudge, a little gentle pull with a small prybar and a soft tap with a hammer might be okay, but think twice before you use these tools. One dent can ruin an otherwise nice job.

5. **Use a persuader.** Cut a 4-in.- to 6-in.-long piece of 1×2. Tap that block with a hammer to move the crown up or down until it's right on the layout lines.

6. **Center your ladder** The piece in the photo is almost identical to the first piece of crown molding I installed. Again, I center the ladder and climb up, carrying the molding in my hands. Until I need them my crown jig is in my pocket, and my nail gun is hooked to my belt.

Install the final pieces

1. **Remove starter piece** Pull the back of the starter piece down and away from the ceiling, then slip it out from the corner.

2. **Grab first piece** Because you left at least 4 ft. or 5 ft. of the first piece loose, it's easy to pull that molding away from the wall. If the last piece is long, slip the far end into a crown hook before climbing the ladder. I positioned the screw in the hook just above the bottom line of the crown, so the two corner pieces wouldn't interfere with each other.

3. **Tack last piece away from corner** It's usually tough to get everything perfect before reaching for a nail gun—carpenters working alone don't have enough hands. Push the crown molding up, check that it's at the layout lines marked with the crown jig, then tack it to the wall a few feet from the corner.

4. **Fit corner** Hold the center of the first piece away from the wall, then press the coped end tightly against the last piece.

5. **Fasten near corner** After making any necessary adjustments—tapping one piece up a little or the other down a hair—fasten the corner to the wall. If you start at the top, the bottom might be open a little until you fasten that edge too.

6. **Press on molding while nailing** Don't rely on the nail gun to pull the cope joint tightly closed. Push on the first piece and press it against the last piece while firing the nail gun.

resources

- **Garrett Wade Co., Inc.**
  Shinto Saw Rasp, Universal
  Squeeze Clamps
  5389 East Provident Drive
  Cincinnati, OH 45246
  800-221-2942
  garrettwade.com

- **Collins Tool Company**
  Collins Coping Foot & Miter Clamps
  P.O. Box 417
  Plain City, OH 43064
  888-838-8988
  614-873-6219
  Fax: 614-873-1676
  collinstool.com

- **Kreg Tool Company**
  Kreg Pocket Hole Jig
  201 Campus Drive
  Huxley, IA 50124
  800-447-8638
  515-597-2234
  kregtool.com

- **McFeely's™**
  McFeely's square-drive screws,
  2P-10 Glue, counter sinks bits,
  bit drivers, prybars, Nail Gripper
  P.O. Box 44976
  Madison, WI 53744-4976
  800-443-7937
  mcfeelys.com

- **Adjustable Clamp Co.**
  Pony Hand Clamps, Miter Saw
  404 N. Armour St.
  Chicago, IL 60622
  312-666-0640
  adjustableclamp.com

- **Brumley Tools, Inc.**
  Trim Gauge
  21520 Mullin
  Warren, MI 48089
  866-278-8665
  trimgauge.com

- **Fein Power Tools**
  MultiMaster cutting and
  sanding tool
  1030 Alcon Street
  Pittsburgh, PA 15220
  800-441-9878
  feinus.com

- **FW Tools, Inc.**
  ThumbSaver nail holder
  119 Southeast Parkway Court
  Suite 240
  Franklin, TN 37064
  615-261-9929
  888-257-9198
  thumbsaver.com

- **LIE-NIELSEN TOOLWORKS**
  low-angle block plane
  PO Box 9
  264 Stirling Road
  Warren, ME 04864
  800-327-2520
  lie-nielsen.com

- **Stabila, Inc.**
  LE 50 Laser Measure, levels,
  lasers, tape measure
  332 Industrial Drive
  PO Box 402
  South Elgin, IL 60177
  800-869-7460
  stabila.com

- **Rockler Woodworking
  and Hardware**
  Tru Angle Protractor
  4365 Willow Drive
  Medina, MN 55340
  800-279-4441
  rockler.com

- **L.S. Starrett**
  Pro-Site Protractors
  121 Crescent St
  Athol, MA 01331
  978-249-3551
  starrett.com

- **TJM Design Corporation (USA)**
  (formerly Tajima Tool Corporation)
  Japanese Pull Saw, caulking gun
  3510 Torrance Boulevard
  Suite 112
  Torrance, CA 90503
  310-540-7900
  888-482-5462
  tajimatool.com

- **Direct Fastener**
  Distributed by Direct Sales Ltd.
  23ga. and 18ga. nail guns
  3605 Commercial St.
  Vancouver, BC VSN 4G1
  CANADA
  604-876-9909
  cadextools.com

- **Rolair Systems**
  air compressor
  606 South Lake Street
  PO Box 346
  Hustisford, WI 53034
  920-349-3281
  rolair.com

index

A

Angles, protractors and, 54, 56, 60, 72, 93
Auxiliary fence, 14

B

Baseboard, 52–69
　bowed walls and, 69
　butt joints, 53
　coping, 56, 66, 67
　cut lists, 53–55, 58–59, 60, 61
　cutting, 25, 55–56, 62–65, 66, 67, 68
　cutting back, 25
　fit problems, solving, 68, 69
　inside corners, 53, 55, 56, 66, 67
　laying out, 53–55 (*see also* cut lists)
　on long walls, 55, 57, 67
　measuring, 53, 58–59, 60–61
　nailing, 57, 67, 69
　outside corners, 53, 57, 60–61, 65, 68
　planning joinery/layout, 53–55
　preparing for casing, 21, 25
　on short walls, 66
　short-point/long-point method for, 53
　tall, cutting, 62–65
　tools and materials, 54
Bevels, 40, 74, 80
Bits, 4, 36

C

Carpenter's square, 4
Casing. *See also* Door casing; Window casing
　assembly jigs, 21
　cutting. *See* Cope joints; Coping; Cutting; Miter cuts
　defined, 17
　removing, 17, 28–29, 39
Caulk
　joints, cutting, 28
　for moldings, 72, 82, 83
Chair rail, 70–83
　beveled edges, 80
　cut lists, 77
　cutting, 73–74, 79–81
　fastening to wall, 82–83
　features, 71
　installing, 74, 76–83
　level line, 76
　measuring, 77, 78
　notched and overlapped, 74
　one-piece, 75
　order of installation, 71
　outside corners, 73, 78
　planning, 71–73
　simple design and joinery, illustrated, 74–75
　tools and materials, 72
　two-piece, advantages of, 74
Clamps, 4, 18, 30
Cope joints
　baseboard, 56, 66, 67
　crown molding, 87–88, 97–98, 101
　overlapping, 66, 67
Coping. *See also* Cope joints
　with coping saw, 88, 97
　foot, for jigsaws, 54
　inside corners, 56, 66, 67
　with jig saws, 98
　jigs, 86, 96
　saws, 54, 55, 88, 97
　with utility knives, 66, 67

Cordless drill, 4
Corners. *See Outside corners and Inside corners* references
Countersink bits, 4
Cove molding
　cutting, 73, 79
　installing, 74, 82
Crown clamps, 86
Crown molding, 84–102
　checking corners with protractor, 93
　cope joints, 87–88, 97–98, 101
　coping clamping station for, 96
　cut lists, 88, 93
　cutting, 85, 87, 88, 94–95, 97–98
　fastening corners, 101
　final pieces, 102
　hooks, 89, 100
　inside corners, 91 (*see also* cope joints)
　installing, 88–89, 99–102
　jigs, 85, 86, 90, 99
　laying out, 87–88, 91
　long runs, 89, 91, 93, 99, 100
　measuring, 87–88, 92–93
　nailing, 99, 100, 101, 102
　one person hanging, 89, 100
　outside corners, 89, 91, 99, 100
　perspective on installing, 89
　spacers for cutting, 87
　starting installation, 99
　stop for cutting, 87
　tools and materials, 86
Cutting. *See also* Cope joints; Coping; Miter cuts
　baseboard, 25, 55–56, 62–65, 66, 67, 68
　casing. *See specific casing types*

caulk joints, 28
chair rail, 73–74, 79–81
cove molding, 73, 79
crown molding, 85, 87, 88, 94–95, 97–98
with miter box, 20
outside corners, 65, 68
to scribe lines, 37

D

Door casing, 16–33
assembly jigs, 21
baseboard preparation, 21, 25
cut lists, 24
cutting, 19–20, 26–27
defined, 17
head casing, 20, 24, 27
installation options, 21
installing, 30–31
leg casings, 19, 24, 26
measuring for, 24
miter points, illustrated, 19
overview, 17
preassembling, 21, 32–33
preparing jamb, 28–29
removing old, 17, 28–29
reveals, 30
tools and materials, 18
Dovetail saws (backsaws), 18, 25
Drilling
countersink holes, 4, 5
pocket holes, 8–9, 38
screw holes, 49
Drills, 4

F

5-in-1 tool, 18, 28, 29

G

Glues/gluing, 18, 36, 51

I

Inside corners (baseboard)
coping, 56, 66, 67
illustrated, 53
measuring, 55
Inside corners (crown molding), 91

J

Japanese saws, 72, 81
Jigs
casing assembly, 21
coping, 86, 96
crown molding, 85, 86, 90, 99
pocket hole, 4, 8–9
Jigsaws
coping foot for, 54
coping with, 98
cutting notches with, 81
cutting scribe lines with, 37, 44
illustrated, 36
safety, 57
using, 57

L

Laser levels, 72, 76
Laser measures, 36, 47, 55, 78, 93
Levels, 72, 76
Low-angle block planes, 36, 45

M

Materials needed. *See* Tools and materials
Measuring
baseboard, 53, 55, 58–59, 60–61
chair rail, 77, 78
crown molding, 87–88, 92–93
for door casing, 24
inside corners, 55
laser measures for, 36, 47, 55, 78, 93
long runs, 55
outside corners, 55, 60–61
tape measures for, 4, 47
Miter box, cutting casing with, 20
Miter cuts. *See also* Cutting
aligning/affixing, 32–33
bevel angles, 40, 74, 80
long points (toes), 19, 53
outside corners, 65, 68
short points (heels), 19, 53
short-point/long-point method, 53
upside down and backward, 73, 79, 85
Miter saw, 4
angles, protractors and, 56
auxiliary fence for, 14
basic cutting techniques, 5–7, 13
repetitive stop system, 15
safety, 5
stand. *See* Workstation
Multimasters, 18, 25

N

Nail grippers, 18, 22–23
Nail sets, 18, 23
Nailers, 72, 82
Nails/nailing
avoiding splits, 22
baseboard, 57, 67, 69
crown molding, 99, 100, 101, 102
door casing, 18, 30–31
protecting fingers, 22–23
setting nails, 23

into studs, 69
using hand-driven nails, 22–23

O
Outside corners (baseboard)
  cutting (mitering), 65, 68
  fitting, 57
  illustrated, 53
  measuring, 55, 60–61
Outside corners (chair rail), 73, 78
Outside corners (crown molding),
  89, 91, 99, 100

P
Pin nailers, 72, 82
Planes, 36, 45
Pocket holes
  drilling, 8–9, 38
  jigs, 4, 8–9
Protractors, 54, 56, 60, 72, 93
Prybars, 18, 28–29

R
Rasps, 36, 44, 81
Removing casing, 17, 28–29, 39
Repetitive stop system, 15
Resources, 10
Reveals, 30, 39, 46, 47

S
Safety
  jigsaws, 57
  miter saw, 5
Sandpaper/sanding blocks, 72, 82, 83
Saws. *See specific saw type*
Screws
  predrilling holes for, 49
  for window stools, 36
Scribes/scribing, 36, 37, 42–45
Setting nails, 23
Shinto saw rasps, 36, 44, 81
Short-point/long-point method, 53
Spring clamps, 18, 30
Stud finders, 54, 69, 82

T
Tape measures, 4, 47
Thumb Savers, 18, 23
Tools and materials
  baseboard, 54
  chair rail, 72
  crown molding, 86
  door casing, 18
  window casing, 36
  workstation, 4
Trim Gauges, 18, 30, 80
Tripods, 72, 76

U
Utility knives, 18, 28, 66, 67

W
Window casing, 34–51. *See also*
  Window stools
  aprons, 37, 38, 50, 51
  assembling self-returns, 51
  components, illustrated, 37
  cutting and installing, 38, 48–51
  laying out, 39
  measuring, 47, 50
  removing old, 39
  reveals, 39, 47
  tools and materials, 36
Window stools
  casing components and,
    illustrated, 37
  cutting and installing, 35–38, 40, 42–46
  edge profiles, 37
  fastening apron to, 51
  laying out, 35, 39
  reveals, 39, 46
  scribing, 37, 42–45
  supports, 37, 41
Wire cutters, 18, 29
Workstation, 2–15
  auxiliary fence, 14
  drilling pocket holes, 8–9
  importance of, 3, 5
  installing supports, 10–11
  installing top, 12–13
  measuring and cutting
    components, 6–7
  parts of, 5
  plan, illustrated, 3
  repetitive stop system, 15
  stop for cutting crown, 87
  supports, 5, 10–11
  tools and materials, 4